MOLECUL

EMBRYOLOGY

MOLECULAR EMBRYOLOGY

HOW MOLECULES GIVE BIRTH TO ANIMALS

J. MICHAEL BARRY

DEPARTMENT OF PLANT SCIENCES
OXFORD UNIVERSITY
OXFORD, UNITED KINGDOM

CRC Press
Taylor & Francis Group
Boca Raton London New York

CRC Press is an imprint of the
Taylor & Francis Group, an **informa** business

CONTENTS

INTRODUCTION

When a sperm penetrates a human egg cell, an embryo results which, within nine months, becomes transformed into a baby of complex shape, composed of millions of cells of varying structure. This book describes experiments, performed over the last 150 years, which have led to our present, and rapidly increasing, knowledge of how molecules interact to cause such remarkable development in animals.

Although there are several excellent advanced texts on developmental biology, it is difficult for beginners to penetrate their wealth of detail and complex terminology. In this book, I have tried to give an accurate summary of the subject that will provide a helpful introduction for students of biology, and will also interest others who want to understand recent advances in this largely undiscovered territory. The material should be easily comprehensible to readers who know only the elements of molecular biology, and the glossary explains to them unfamiliar terms and concepts.

CHAPTER ONE

GROWTH OF AN EMBRYO IS FOUNDED ON REPEATED CELL DIVISION

DOES A VITAL FORCE DIRECT ANIMAL DEVELOPMENT?

The development of an animal begins when a sperm, which is essentially a mobile package of male genes, fertilizes an egg. The egg contains not only the female genes, but also a complex assembly of **molecules** needed to initiate development. The transformation of the single cell of the fertilized egg into an adult animal, usually with millions of cells of many different kinds all located in the correct position, is a complex process. Can all these precise changes result from the interaction of the component molecules of the egg according to the normal laws of chemistry? Until recently, many biologists had thought that the answer is no. They believed that an additional vital force must operate in living cells. It is worthwhile first to consider the theories of these "vitalists," as they are fundamentally related to the theme of this book.

Theories of vitalists that could conceivably be tested by experiment are of two distinct kinds. The first can be illustrated by the ideas of Justus von Liebig,[1] a famous German biochemist of the 19th century. In his textbook, *Animal Chemistry,* he appears to agree with the "mechanists" that a living organism results from its component molecules being correctly positioned and that a living organism could, at least in theory, be made in the laboratory. But he believed that when the component molecules are in these positions, a new force of nature manifests itself. This force, he believed, is inherent in all molecules but acts only in the complex structure of the cell. Thus, research in biochemistry would reveal the laws of its action just as experiments in physics revealed the laws of magnetic attraction.

The possibility of a vital force was still entertained in 1945 by the famous physicist, Erwin Schrödinger in *What Is Life*?[2,3] This book made a great impact on physicists and persuaded some to enter biology in the hope of discovering new natural laws. Schrödinger[2] concluded, "living matter, while not eluding the 'laws of physics' as established up to date, is likely to involve 'other laws of physics' hitherto unknown, which, however, once they have been revealed, will form just as integral a part of this science as the former" (p. 73). This conclusion was based on his opinion that genes display an inexplicable stability. Schrödinger[2] had correctly deduced that a gene is whole or part of a single large molecule, but incorrectly (as we now realize) concluded that these single molecules have a "durability or permanence that borders on the miraculous." He mentioned the gene responsible for the famous lip of the Hapsburg dynasty. "How are we to understand that it has remained unperturbed by the disordering tendency of the heat motion for centuries?" Schrödinger[2] compares the biologist to an engineer who is familiar with steam engines but encounters an electric motor for the first time. Although it is made of familiar materials, he suspects that laws that he does not understand are involved in its function. The biologist nevertheless believes that these are comprehensible ordered laws. "He will not suspect that an electric motor is driven by a ghost because it is set spinning by the turn of a switch, without boiler or steam" (p. 81).

In contrast, vitalists of the second theory did believe in a ghostly force and that the basis of life lies outside the normal realm of science. Their theories were varied, but the most clearly defined were those of Hans Driesch,[4] who died in 1942. He started his career as an embryologist and made fundamental contributions to the subject. But he became convinced that animal development would never be explained solely by the interaction of forces of nature of the kind typically investigated by scientists. As a result, he retired from experimental biology in the early 1900s and became a professor of philosophy at the University of Heidelberg, where he developed his theories of vitalism.

Driesch[4] believed that living organisms differ sharply from nonliving organisms in the possession of a vital force unlike any force familiar to scientists—a purposeful directive force like that suggested by Aristotle. Its purpose is that a living organism should grow to maturity and reproduce. It acts by directing the component parts of the organism along paths they would not take under the sole influence of the normal forces of nature, so causing the organism to develop and function correctly. Driesch claimed that the existence of this force was proved beyond all reasonable doubt by certain experiments in embryology.

For example, when the fertilized egg of a sea urchin develops into an adult, it divides first into two cells; these cells each divide into two more to produce four, and each of these four cells by further division give rise to different parts of the adult. But if the four cells are broken apart and allowed to develop separately, they each form, not different fragments of the adult, but a complete adult sea urchin. Driesch concluded that the four cells are identical and that the normal forces of nature could not cause different parts of an adult to arise from each of four identical cells. (His view has been expressed succinctly by the famous biologist Ernst Mayr,[5] "What machine, if cut in half, could function normally?") Again, if the limbs of certain Amphibia are removed, perfect new limbs develop in their place. Driesch considered that this could come about only under the influence of a directive vital force that is attempting to resist damage to the organism and that this could never occur if the component molecules were merely under the sway of the blind forces of nature of the inanimate world. The action of Driesch's vital force requires that the normal laws of nature can be violated in living organisms, and Driesch considered in detail how this violation might be detected.

Even the most famous experimental embryologist of all time, Hans Spemann, a contemporary of Driesch, was undecided about vitalism. As we shall see, Spemann's experiments form the foundation of our understanding of embryology, and for this work he was awarded the Nobel Prize in 1935. In an address in 1923, he spoke cautiously in relation to Driesch's work, "I would leave it undecided whether, and how soon, we will encounter ultimately unsolvable problems in the direction taken by Driesch."[6] Nevertheless, Spemann's ideas, like those of most of his contemporaries who were influenced by the impact of the German romantic movement on science, had overtones of vitalism. Thus, in 1938, at the end of a book in English summarizing his life's work, Spemann[7] wrote rather obscurely, "these processes of development, like all vital processes, are comparable, in the way they are connected, to nothing we know in such a degree as to those vital processes of which we have the most intimate knowledge, viz., the psychical ones.... Here and there this intuition is dawning at present. On the way to the high new goal I hope I have made a few steps with these experiments."

Both vitalist theories can be tested by trying to discover, as biologists do, whether living organisms can be completely explained by the interaction of their component parts according to the familiar laws of chemistry and physics. Over the last 50 years, molecular biologists have shown that most fundamental processes of living cells are founded on the interaction of large molecules according to the normal laws of chemistry. No new natural force of the kind suggested by Liebig or Schrödinger appears to act. But until very recently, there was little experimental evidence that a directive vital force of the kind suggested by Driesch is not required during the development of a fertilized egg into an adult. This evidence is now being provided by the experiments described in this book. As a result, the brain is becoming the last region of the body where it is not entirely unreasonable to suggest that a vital force operates.

THE DIVISION OF EMBRYONIC CELLS

Animals are composed of cells that increase in number by the division of one cell into two (mitosis). Animal development is the transition from the single cell of the fertilized egg to a multicellular adult; hence, cell division is fundamentally involved in the developmental process. This chapter describes some features of cell division that are important in understanding animal development and illustrates these features with details of cell division in the embryos that are most commonly used for research.

The microscopic structure of an animal cell when not dividing is illustrated in **Figure 1-1**. It is enclosed by the cell membrane. In the center of the cell is the nucleus, which contains the thread-like chromosomes that are largely invisible until cell division. These chromosomes occur in **homologous pairs**, that is, of identical shape, with the members of each pair being derived from the two parents at fertilization of the egg. The nucleus is enclosed by two closely aligned membranes, and the outer one branches through the **cytoplasm** (viscous fluid portion of the cell) to form a system of blind tubes, named the **endoplasmic reticulum**. Outside the endoplasmic reticulum lie the **ribosomes**, particles that are involved in the formation of proteins. The double nuclear membrane is pierced at many points by nuclear pores, each

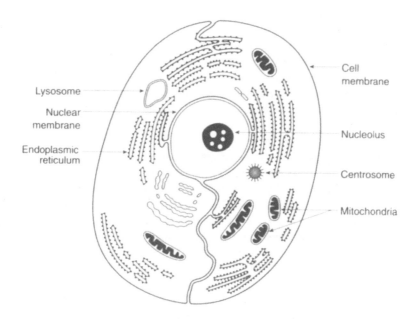

FIGURE 1-1 Structure of typical nondividing animal cell as revealed by the electron microscope.

surrounded by a pore complex of precise structure. The pores enable all but the largest molecules to pass from within the nucleus to the cytoplasm. The cell contains several distinct structures, including nucleoli (usually two) within the nucleus, a centrosome, and many mitochondria and lysosomes in the cytoplasm—all with separate functions. Also, numerous protein enzymes direct the conversion of one kind of molecule into another. Some enzymes are common to all cells, whereas others are peculiar to a particular type of cell.

Not shown in the diagram is the **cytoskeleton**, whose filaments ramify throughout the cell and are attached to certain structures. They are important in directing cell division and in causing changes in the movement and shape of cells, which occur repeatedly during animal development. The three kinds of filaments are actin filaments and microtubules, which are continuously degraded and reformed, and the more stable intermediate filaments. **Actin filaments** are built up within the cell by linking molecules of the protein actin into long chains, which are one molecule thick but coiled into a helix (corkscrew). When associated with filaments composed of the protein myosin, long arrays of actin filaments can contract in length, as they do in muscle cells. A dense network of actin filaments, associated with other proteins, lies beneath the cell membrane to form the **cortex**, which gives strength to the cell and directs changes in its shape and movement. Actin filaments also branch throughout the cytoplasm. **Microtubules** are formed by linking molecules of the protein tubulin into long **protofilaments**; 13 of these protofilaments are aligned side by side into a hollow tube to form the microtubule. The microtubules are less plentiful than actin filaments and are formed in the **centrosome** (see Fig. 1-1) beside the nucleus and radiate from it. **Intermediate filaments**, which are composed of fibrous proteins, largely extend from around the nucleus to the cell membrane and appear to stabilize the cell.

The essential requirement of most cell divisions is to produce two daughter cells with the same number of chromosomes (and hence the same number of genes) as the parent cell. (The exception is the "reduction division," which occurs during the formation of egg and sperm cells, in which the members of each homologous pair of chromosomes separate to give rise to two cells with half the usual number of chromosomes.) The movements of cell division are controlled by the cytoskeleton, which also determines the relative sizes of the daughter cells and the distribution of cell components between them. The structure of a dividing cell is outlined in **Figure 1-2**.

Division is preceded by the nuclear membrane disintegrating and the chromosomes becoming condensed and visible. Each chromosome is seen to be composed of two closely adhering identical chromosomes (**chromatids**), which result from an earlier self-copy of each chromosome that was received from the parent cell. At the same time, the two centrosomes, which have formed by duplication of the original one, migrate to opposite ends of the cell

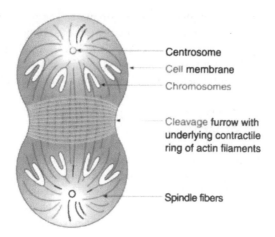

Centrosome
Cell membrane
Chromosomes

Cleavage furrow with
underlying contractile
ring of actin filaments

Spindle fibers

FIGURE 1-2 Diagram of dividing cell.

with a "spindle" of microtubules between them. The members of each pair of chromatids (not to be confused with the homologous pairs of chromosomes) become attached to the spindle fibers and are drawn apart to opposite sides of the cell by their contraction. A ring of actin filaments, which has formed beneath the cell surface, then contracts, to form a circular depression (its **cleavage furrow**), which finally closes, giving rise to two separate cells. In each new cell, the nuclear membrane reforms and the chromosomes again become invisible. During the interval before the next cell division, each chromosome is duplicated to again form two identical chromosomes that will separate from one another at the next division. Also during this interval, if the next division is to give rise to two cells identical with the parent cell, every other component of the cell must double in quantity.

During animal development, a parent cell must often give rise to daughter cells, which, although having chromosomes similar to itself, differ from it in size and molecular composition. The size, composition, and position of daughter cells is determined by the position of the contractile ring. This always forms halfway between the centrosomes, whose position appears to be determined by actin filaments of the cell cortex. The period from the formation of a new cell by the division of its parent cell until the division of this new cell into two cells is known as the **cell cycle**. The following are the different types of cell cycles that occur in developing animals.

1. A cycle in which the cell *doubles in weight*, by each of its components doubling in quantity, followed by equal division of the cell contents to form two daughter cells identical with the original parent cell. Liver cells

multiply in this way late in development after they have acquired their final characteristic structure.

2. A cycle in which the cell again increases in weight before division, but *also changes in molecular composition*. It then divides to give rise to two identical cells that differ from the original parent cell.

3. A cycle in which the cell again increases in weight, with or without change in molecular composition, but in which *division is asymmetric*. The cell components are unevenly divided to form two cells whose composition differs both from one another and from that of the original parent cell.

4. A cycle that is similar to #3 except that only *one of the daughter cells differs from the original parent cell,* thus retaining the original cell population. The dividing cell is known as a **stem cell**.

5. A cycle that gives rise to two identical cells, but in which *division is not preceded by any increase in cell weight,* although new chromosomes and cell and nuclear membranes are formed at the expense of other components. The daughter cells have half the volume of the original parent cell and a higher ratio of cell surface to cell contents. The first division of the fertilized egg of many animals is of this kind.

6. A cell cycle that is similar to #5 in that there is *no increase in cell weight*, but in which *division is asymmetric*, forming two cells of different, rather than identical, composition. Such cycles occur early in the development of many animals.

7. A truncated cycle in which the *cell changes in composition but does not divide*. Such cycles occur toward the end of development to form specialized cells that will never divide again, such as red blood corpuscles that are packed with the protein hemoglobin that transports oxygen.

CELL FATE AND CELL LINEAGE

One of the first requirements in studying development is to bring some order into our thinking about cell division, and this has required the invention of some jargon. For example, a population of cells formed by division of a single cell, and further division of its progeny cells, is called a **clone**. Thus, all cells of an adult animal are a clone descended from the fertilized egg; roughly half of these cells are a clone descended from one cell of the two-celled embryo, roughly one-fourth of these cells are a clone from one cell of the four-celled embryo, and so on. The smallest possible clone within an animal is two cells derived from the same parent cell. These smaller clones are called **subclones**. Cells of a subclone may or may not be adjacent to one another in the adult, because many cells migrate away from one another after their formation.

Every cell of a developing animal is said to have a **cell fate**, which is to give rise to a particular clone of progeny cells, although in practice these cannot always be identified. Thus, the fate of the fertilized egg is to produce every cell of the embryo at any stage of development; the fate of a primary spermatocyte (i.e., sperm precursor cell) is to produce four mature sperm cells; the fate of an immature red blood cell is to produce one mature red blood cell. Conversely, each cell of an embryo or adult has a precise **cell lineage** (ancestry or pedigree). The lineage of each cell traces back to the fertilized egg through a series of cells—first to a single parent cell, then to a single grandparent cell, then to a single great-grandparent cell, and so on.

The latter terminology clarifies our thinking about cell division and leads us toward making precise studies of cell fate and cell lineage, which have provided important clues to the mechanism of development. Various methods of study are used, the simplest being to study whole embryos or sections of embryos under the microscope at various stages of development.[8] This method leads to the discovery of which cells give rise to which tissues. Another possible method is to destroy groups of cells in an embryo by a laser beam and search for missing tissues at a later stage of development. Also, a single cell, or a group of cells, can be labeled and their descendant cells can be identified at a later stage. Cells on the surface of the embryo can be labeled by staining, but the stain becomes invisible after a few cell divisions. More successfully, certain fluorescent dyes or enzymes can be injected into single cells from a fine glass needle under a microscope, and the descendant cells can be identified by fluorescence or by the color produced by action of the enzyme on an added reagent.

CELL DIVISION IN THE THREADWORM

In most animals, the sequence of cell divisions is slightly different from one individual to the next. This limits the precision of studies of cell fate and lineage. But it is possible to imagine a species of animal in which, at each step in development from egg to adult, every individual has the same number of cells of each kind placed in exactly the same relative positions (just as identical houses could be built by always placing at each stage the same number of each kind of brick in the same positions). Fortunately, an animal, *Caenorhabditis elegans,* comes close to this theoretical type and provides a model for studying cell division in development.[9]

C. elegans is a nematode, or threadworm, about 1 mm long, which normally lives in the soil and feeds on microorganisms. A population can be kept alive indefinitely by storage in glycerol in liquid nitrogen and, when required, can be allowed to grow and reproduce at 20° C on cultures of *Escherichia coli* bacteria. The adult *C. elegans* has a long symmetric body made up of familiar tissues such as muscles, nerves, and an outer skin (**Fig. 1-3**). It is transparent throughout development and cell nuclei can be seen in living animals under a microscope. The fertilized egg develops into the adult in two stages. By about 13 hours after fertilization, it has developed into a small worm with 556 cells in the male. It then breaks out of the egg to begin development into the adult. Every adult male has exactly 1031 cells (apart from around 1000 sperm cells), and these cells are placed in the same relative positions in every individual, as are the cells at every stage of development up to the adult. (To be strictly accurate, the precise number refers to cell nuclei rather than cells, since the nuclei are occasionally not separated by cell membranes.)

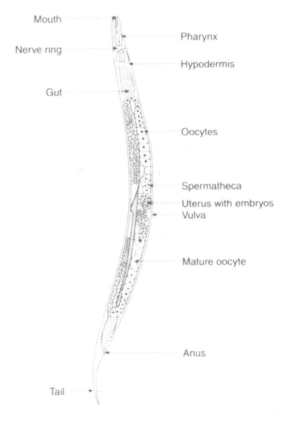

FIGURE 1-3 Section through adult *Caenorhabditis elegans.*

Its exact cell structure at each stage of development has been painstakingly deduced over a number of years.[10] Each embryo was put on gel on a microscope slide and covered with a glass coverslip. After hatching, the worm was fed under the coverslip with a suspension of *E. coli* bacteria and could crawl on the gel surface. It was possible to record at each stage the positions taken up by the progeny of each dividing nucleus, and, except for sperm cell nuclei, these positions are identical in every individual. Observations were recorded by drawings and photographs and by time-lapse video recording to give a moving picture of development. They were supported by photographs of sections of the animal taken under the electron microscope.

Figure 1-4 shows two parts of the lineage of all cells that arise during development of the adult *C. elegans*. Since, at each stage of development, all *C. elegans* individuals have identical numbers of each kind of cell and identically positioned cells, a particular code of letters can be assigned to each cell at each stage for its identification. Hence, the fate of any cell can be precisely specified by the codes of its progeny cells and its lineage by the codes of its ancestors. These are virtually the same in every individual. The cells formed in the first few divisions after fertilization are denoted by code letters that had historical significance in studies of nematode development: P_1, P_2, P_3, and P_4 for cells on the germ line, and AB, EMSt, MSt, E, C, and D for the remainder. For the cells formed thereafter, a system of letters is used that denotes the position of a cell, relative to its sister cell, immediately after the division that produced it. Therefore, a and p denote anterior and posterior; d and v, dorsal and ventral (i.e. upper and lower); and l and r, left and right. Thus, a cell ABpra, which is present for a short time in early development, is the anterior daughter of the right-hand daughter of the posterior daughter of AB. Hence, the lineage of this cell is given by the codes of its ancestral cells back to the fertilized egg, and its fate in the adult is given by the codes of all its progeny cells within the adult.

As previously mentioned, most of the different types of division that cells can undergo occur in the lineage of *C. elegans*. After fertilization, the egg cell divides without previous growth to give rise to cells (AB and P_1) that differ from one another and from the parent cell. There are points in the sperm cell lineage (not shown in Fig. 1-4) at which a cell gives rise to a clone identical with itself by a series of cell cycles in which each cell doubles in size and divides into two cells identical with the original, thereby maintaining the population of sperm precursor cells. Also found are cycles that yield two identical cells that differ from the original parent cell, and cycles that yield one cell identical with that of the parent and one different.

Certain important conclusions can be drawn about the development of *C. elegans* from its lineage. First, there is no simple underlying plan. Suggestions have previously been made that all adult cells of a particular type,

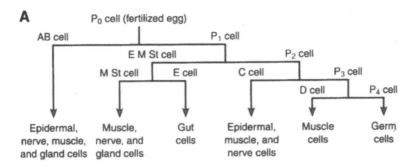

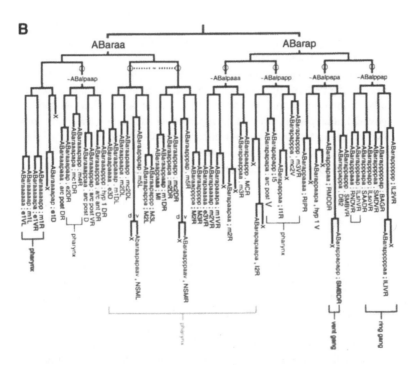

FIGURE 1-4 A: The early divisions of the fertilized egg of *Caenorhabditis elegans*. **B:** Part of the complete sequence of cell divisions. (Adapted from Sulston JE, Schierenberg E, White JG, Thomson JN. The embryonic cell lineage of the nematode *Caenorhabditis elegans*. *Dev Biol* 1983;100:64–119.)

such as nerve cells, are a clone derived from a single ancestor, but this is not true for this nematode. All sperm cells and all intestinal cells in the lineage are in fact members of clones that include no other cell type, but all other kinds of cells arise on various branches of the lineage together with cells of a different kind. Some branches even end in a division that yields two different cells such as muscle and nerve. This has important implications for understanding the molecular basis of development: namely, the theory is excluded that genes peculiar to a certain cell type remain potentially active only in one branch of the lineage tree. Other interesting features include programmed cell death, **apoptosis**, in which certain cells always die before development is complete—this being part of the mechanism for giving the correct body structure. Also, the time taken to complete a cell cycle is constant within each main branch of the lineage but differs from one main branch to another, a point of interest to those who try to understand the mechanisms that control cell division.

Such studies of an obscure worm might be thought to have little relevance to human development, but this view has little justification. About 50 years ago, to the surprise of biochemists, the molecular interactions that control the energy production and inheritance of living cells were found to be remarkably similar from bacteria to humans. It was concluded that the still-undiscovered molecular interactions that control animal development would be found to be similar from one animal to the next. As these interactions began to be uncovered, unexpected differences appeared whose significance was unclear. However, many fundamental molecular mechanisms are now being found to be remarkably similar among animals separated by millions of years of evolution. Clearly, any animal that is especially suited for experimental study should be studied.

CELL DIVISION IN FRUIT FLIES

Cell division in most animals is less ordered than that in *C. elegans*. Higher animals have many more cells, and adults differ widely in the number of each type of cell and, to some extent, in cell positioning. Hence, each cell at each stage of development cannot even in theory be uniquely identified or assigned an exact fate or lineage. However, less precise information about cell fate can be discovered, which can be useful. What follows are some facts about cell division in three other embryos on which research is being concentrated: fruit flies, frogs, and mice.

The fruit fly, *Drosophila melanogaster*, is a familiar sight around decaying rubbish and is famous for having yielded in the first half of the 20th century intricate details of the mechanism of inheritance. It is easily bred in the laboratory and has a short life cycle. Mutant flies with defective genes are readily produced from parents that have had x-radiation or have been fed mutagenic chemicals, and this has recently enabled important discoveries about the mechanism of development to be made, as described in Chapter 6.

The start of development in *Drosophila*, as in other insects, is different from that in most animals in that the nucleus of the fertilized egg proliferates, but no new cells are formed (**Fig. 1-5**). After fertilization, the sperm nucleus unites as usual with the egg nucleus to form the single nucleus of the fertilized egg, which then divides to give rise to two daughter nuclei. However, these do not become surrounded by cell membranes to give rise to separate cells as in most animals. Instead, the cycle of nuclear division (always preceded by the replication of each chromosome) is repeated seven more times in synchrony to give 256 (i.e., 2^7) nuclei distributed through the cytoplasm of the egg cell. A few of these nuclei give rise to a cluster of about 40 **pole cells** at the posterior of the egg, which are the ancestors of the adult's egg or sperm cells. But most of the nuclei migrate to the surface of the egg where, after further multiplication, they form a single layer beneath the cell membrane. After only 3 hours from fertilization, membranes form around each of these 6000 or so nuclei to form a single layer of cells (the **blastoderm**) beneath the whole of the surface of the egg. Development then proceeds rapidly by cell division and differentiation: One day after fertilization, a maggot-like larva hatches from the egg. It then feeds and grows and, after about 4 more days, secretes a cuticle to form the inert pupa (see Fig. 1-5) within which the adult fly develops and emerges a few days later.

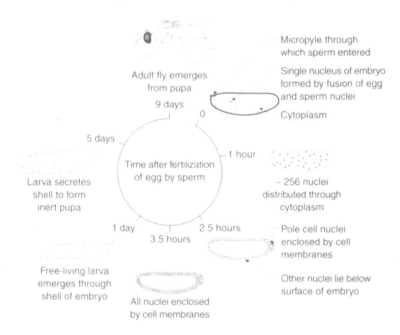

FIGURE 1-5 Life cycle of *Drosophila melanogaster*.

Studies of cell lineage and cell fate in *Drosophila* are influenced by the fact that development to the adult occurs in two stages: (1) from egg to pupa and (2) from pupa to adult fly. Studies of cell fate[11] show that development of the larva from the single layer of blastoderm cells of the early embryo is precisely ordered (**Fig. 1-6**); cells that are adjacent in the blastoderm tend to give rise to progeny cells, which lie near one another in the larva. The blastoderm first undergoes **gastrulation** by folding inward to form two cell layers that lie close to one another: the **ectoderm** (outer) and the **endoderm** (inner). Between these layers is formed a third layer, the **mesoderm**. From these three layers of cells, different organs of the larva develop. The important result of gastrulation is this: cells that are to form the different body tissues are brought roughly into the positions where these tissues should be formed.

Most of the larva disintegrates within the pupa, and the epidermis of the adult fly, as shown by fate mapping, is formed from small reserves of larval epidermal cells called **imaginal discs (Fig. 1-7)** and **histoblast nests**. Muscle, nerve, and other tissues are similarly formed from undifferentiated cells of the larva. It is as if two different animals, the larva and the fly, develop from the same fertilized egg. Each imaginal disc is composed of a layer of similar-appearing cells in the form of a collapsed balloon. One disc, the genital disc, lies in the midline of the larva and develops into the sex organs of the fly. There are 10 other kinds of discs, which occur in pairs on either side of the larva and are named according to the part of the adult that they form: eye—antennal disc, first leg disc, wing disc, and so on. The cells of the histoblast nests divide to form the epidermis of the abdomen. Each of the regions formed abuts another like pieces of a jigsaw puzzle. Groups of about 20 cells, which each give rise to an imaginal disc, can first be identified before the larva hatches from the egg. Within the larva, the cells of each disc increase to about 50,000.

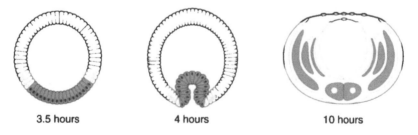

| 3.5 hours | 4 hours | 10 hours |

FIGURE 1-6 Sections through larvae of *Drosophila* at different times after fertilization. Fate mapping proves that cells on ventral surface at 3.5 hours (in light blue) invaginate at gastrulation (4 hours) and later form mesoderm cells.

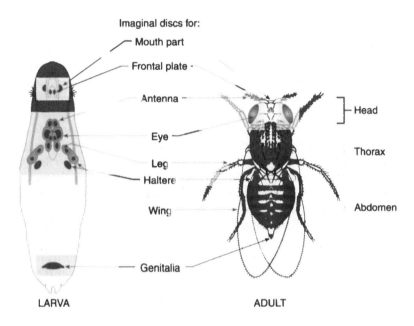

Imaginal discs for:

Mouth part

Frontal plate

Antenna

Eye

Leg

Haltere

Wing

Genitalia

Head

Thorax

Abdomen

LARVA

ADULT

FIGURE 1-7 Principal imaginal discs of *Drosophila* larva and the adult structures that they form.

Surprising results, the significance of which is still unclear, have come from studies of the fate of individual cells of imaginal discs of the *Drosophila*.[12] It is possible to induce a change in chromosome structure within one cell of a larva, which will enable all progeny of this cell to be identified by, for example, abnormal bristle color in the epidermis of the adult. The gene that gives rise to this trait must be recessive, and the flies must be heterozygous. That is, they must carry one gene for the normal trait and another for the recessive trait and hence reveal only the normal trait. Larvae of these flies are exposed to x-rays to induce **mitotic recombination**. This is a rare event, with not more than one cell normally being transformed per individual. It involves mutual exchange of segments between chromosomes before cell division, which can give rise to a single cell that is homozygous, that is, carrying two genes for the recessive trait. Hence, its progeny cells will show the trait. If mitotic recombination occurs in an imaginal disc, the clone of transformed cells in the resulting adult fly is seen as a compact group of marked cells unmixed with neighboring cells.

Attention was concentrated on adult epidermis, on and near one wing, which is formed from one of the pair of wing discs. If larvae were irradiated immediately after hatching, a marked clone occupied about 1/20 of this epidermis, consistent with microscopic evidence that there are about 20 precursor cells in the disc at the time of irradiation. However, a surprising fact was observed about the distribution of these clones: they are restricted to one side or the other of an invisible demarcation line across the part of the body surface formed from the wing disc (**Fig. 1-8**). This would not be surprising if the two parts formed discrete structures, but they do not. The edge of a clone is normally irregular, but if it happens to fall along this demarcation line, its edge is straight. There is no sign that a septum (dividing wall) or barrier prevents cells from mixing, and the demarcation line arbitrarily crosses structures such as wing veins. The line appears to result from cells on each side differing in surface properties, which make like cells adhere and repel unlike cells, since individual cells isolated from different sides of the line segregate from one another after mixing. Irradiation of older larvae suggested that, as cell division proceeds, the two **compartments** in the wing disc are further divided and subdivided into smaller compartments. Similar compartments were found in epidermis formed from other imaginal discs.

Many years ago, a mutation in a gene named *engrailed* had been found to abolish differences between the front and rear of the wing, with the rear now being a mirror image of the front. It was as if the front and rear of the wing disc now respond identically rather than differently to some inducing agent that diffuses in both directions from a line across its center. When marked clones were induced in embryos of these mutant flies, they were not restricted to one of two compartments and could straddle the normal demarcation line. This suggests that the difference in surface properties had been abolished. Clones in an otherwise normal embryo that had been both visibly marked and made homozygous for the *engrailed* mutation could spread from the rear to the front compartment but not from the front to the rear. This suggests that the clones now had surface properties of cells at the front, that is, that the *engrailed* gene is normally active only in the rear compartment.

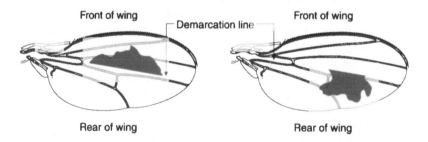

FIGURE 1-8 Clones of marked cells (in red) do not cross imaginary straight demarcation line across wing of *Drosophila*.

After all these facts had been discovered, around 1970, a comprehensive mosaic theory of animal development was constructed from them: as cell division proceeds, the cells throughout the embryo are successively divided and subdivided into compartments by the activation at each step of a particular "selector" gene (such as *engrailed*) in a subgroup of cells, and the compartments subsequently develop as partially independent units. Each active selector gene would produce two changes. It would cause the cells (1) to adhere to form a compartment and (2) to render those genes that are needed to form the characteristic structures of the compartment accessible to inducing signals that arise later. Subsequent experiments show that the whole of animal development is certainly not based on this mechanism, although certain parts of an insect's body may be determined in this way and possibly also parts of the hindbrain of mammals. Other work relating to compartments is described in Chapter 7.

CELL DIVISION IN FROGS

The development of frog embryos has been studied in detail since the 19th century.[13] The favorite species is now *Xenopus laevis*, the South African clawed toad—in fact, a frog. These animals are easily kept in tanks of water and need to be fed compounded pellets only three times a week, followed by a change of water. By injection of chorionic gonadotropin, mature females can be induced to lay eggs every few months, and males can be induced to fertilize the eggs by injection of the same hormone. Small quantities of experimental solutions can easily be injected into the eggs, which are about 1.3 mm in diameter, and into the cells of the early embryo, from a fine glass needle held in a micromanipulator. Because the cells have supplies of nutrients in the form of yolk, small fragments from early embryos continue to develop in salt solutions.

Egg cells of *Xenopus*, like those of other frogs, are surrounded by a protective transparent **vitelline membrane** and are embedded in gel. The top half of each egg as it floats in water is visibly different from the bottom, and each gives rise to different regions of the body. The bottom half, known by the antique name **vegetal hemisphere**, becomes the rear of the early embryo. It is pale yellow and its cytoplasm is packed with yolk granules that nourish the embryo until, as a tadpole, it hatches by breaking through the egg membranes and is able to feed. The top half of the egg, the **animal hemisphere**, becomes the front of the early embryo. It has black or dark brown pigment grains lying near the surface, under which lies the egg nucleus, and its cytoplasm has fewer yolk granules than does the vegetal hemisphere.

Although the upper and lower halves of the *Xenopus* egg are visibly different and form different parts of the body, the unfertilized egg is radially symmetric about an imaginary line passing from the **animal pole** (the center of the animal hemisphere) to the **vegetal pole**, meaning that bisection along this line always produces identical halves. This symmetry is broken by entry of the sperm, which occurs within the animal hemisphere toward the equator of the egg and leaves a dark **sperm entry point** on the egg surface (**Fig. 1-9**). The sperm entry point lies on the lower side of the early embryo and determines the dorsal-ventral (upper-lower) axis of the embryo (but not that of the adult because of later cell movements), since the plane of the first division and the ventral midline of the embryo will pass through this point. Fertilization is soon followed by rotation of the outside of the egg relative to the inside. It is as if the egg is composed of two rigid units: the cell membrane plus underlying cortex with its network of actin filaments and a large spherical core. These units rotate about 30 degrees relative to one another, with the animal pole moving toward the sperm entry point. This rotation is essential for correct subsequent development.

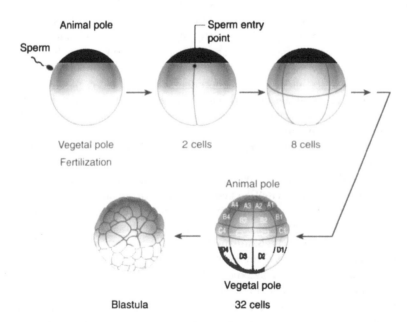

FIGURE 1-9 Cleavage divisions in early embryo of *Xenopus*. In embryo with 32 cells, the position of cells is still almost identical in every embryo. Hence, cells can be identified by code numbers as shown.

Ninety minutes after entry of the sperm, the egg divides into two cells along a plane that passes through the animal and vegetal poles and also through the sperm entry point. The progeny of these cells form the right and left halves of the early embryo. After another 35 minutes, these cells divide again. The cleavage plane again passes through the animal and vegetal poles but at right angles to the first to give rise to the upper and lower halves of the embryo. After another 35 minutes, these four cells divide into eight along a plane that passes through the equator of the egg between the animal and vegetal poles to separate the front and rear of the embryo. These eight cells undergo eight more simultaneous cycles of division at 35-minute intervals to give rise to exactly (or almost exactly) 2048 cells grouped into a spherical **blastula** with a cavity, the **blastocoel**, beneath the animal pole.

Cell divisions up to this stage are called **cleavage divisions**, because they subdivide the preexisting spherical mass of cytoplasm without growth of the embryo. Until at least eight cycles of division are completed, the number and position of cells in every embryo are almost identical. Hence, the cells at each stage can be identified by code numbers as in *C. elegans* (see Fig. 1-9). After 11 divisions synchrony ends, the cell cycle lengthens, and the number, position, and size of cells come to differ among individuals. **Invagination** (the inward movement of cells) of the blastula follows, with further cell division to form a gastrula with about 19,000 cells. Finally, after further development at the expense of the yolk, the tadpole breaks out of the egg at about 4 days after fertilization to grow and "metamorphose" into the adult frog.

The fate of groups of cells of *Xenopus* embryos has been studied by staining. Gel impregnated with the stain is pressed against a particular group of cells on the surface of an early embryo, which, at a later stage, is fixed and sectioned and the location of the stained cells discovered. The cells are mostly found to lie adjacent to one another and in roughly the same position from one embryo to the next; therefore, they have a more or less precise fate. The fates of single cells have also been studied by injecting into them the enzyme peroxidase or a fluorescent compound. Progeny cells are identified in microscopic sections of later embryos by a staining reaction for the enzyme or by fluorescence. The progeny of any single cell of the 32- or 64-celled blastula are found to lie adjacent to one another in later blastulae. In later embryos, most give rise to the types of progeny cells predicted by other studies of cell fate and in the same locations. However, some are found scattered in different tissues as different cell types.

Fate maps, such as that in **Figure 1-10**, reflect the extensive movements of sheets of cells that begin at gastrulation and are largely responsible for shaping the body. Gastrulation involves the movement of cells of the

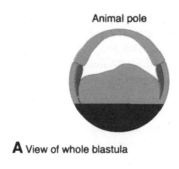

A View of whole blastula

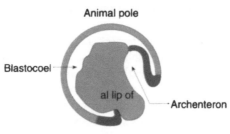

B Section of mid gastrula

FIGURE 1-10 Fate (within gastrula) of different regions in *Xenopus* blastula. Ectoderm shown in light blue, mesoderm in dark blue, and endoderm in gray.

vegetal hemisphere into the interior and the stretching of the animal hemisphere to cover the entire embryo. **Invagination** starts at the **blastopore**, an indentation below the equator opposite the sperm entry point. Invagination is most extensive at the **dorsal lip** of this indentation, where the invaginating cells move forward beneath the animal hemisphere to form a new cavity, the **archenteron**, within the blastula—just as a finger pressed into an inflated balloon forms a new cavity. The blastopore gradually becomes circular and moves down to the vegetal pole to eventually form the anus. The progeny of cells of the animal hemisphere of the blastula largely become ectoderm cells, including cells of the epidermis, which covers the outer surface of the embryo, and nerve cells. Cells of the vegetal hemisphere move inward to give rise to endoderm cells, which include cells of the gut and lungs. A belt of cells at the equator also moves inward and toward the future front of the animal to form mesoderm cells, including blood vessels, muscle, and various glands. Soon after gastrulation, the ectoderm on the dorsal side folds to form the **neural tube** from which the brain and spinal cord develop.

CELL DIVISION IN MOUSE EMBRYOS

Money is needed for research in embryology, and this money comes most easily to researchers whose results could increase our understanding of human development. However, since experiments on humans are largely impossible, other mammals must be studied. The usual choice is mice, since they are easy to rear and have a short gestation period of about 21 days (from insemination to birth). In addition, techniques for experimenting on mice have recently been revolutionized as described in Chapters 5 and 7.

Eggs of mice, like those of other mammals, are small and almost without yolk, since nutrients are provided from the mother's bloodstream. The egg is about 0.1 mm in diameter, and the cell membrane is surrounded by a vitelline membrane as in frogs, which is further surrounded by a layer of protein (the **zona pellucida**). After mating, eggs are shed from the ovary into the oviduct, where they are fertilized and then pass down into the uterus. There, about 4 days after fertilization, the egg sheds its outer protein layer and attaches to the wall of the uterus, where the remainder of its development occurs. The fertilized egg gives rise to both the embryo itself and to extraembryonic tissues such as the placenta, which enables the exchange of nutrients and excreta between the blood of the mother and embryo and is voided as the afterbirth.

As in other mammals, cell division in the early embryo is infrequent. The egg first divides about 24 hours after fertilization, and subsequent divisions occur at intervals of about 12 hours (**Fig. 1-11**). The first three cleavages divide the fertilized egg into eight apparently identical cells, which have only a small part of their surface in contact with one another. These eight cells then **compact**; that is, they become pressed together and difficult to distinguish. By 3.5 days, the embryo has about 64 cells, about 48 of which form an outer layer one cell thick (the **trophectoderm**), which encloses a cavity filled with fluid (the blastocoel). The remaining cells form a cluster (the **inner cell mass**) on one side of this cavity.

Knowledge of cell fate in mice is limited, because development is hidden within the mother and experiments are difficult to perform. Early work was confined to examining sections of embryos of increasing age under the microscope. But recent experiments have been made easier by techniques for removing early embryos from mice, maintaining them for several days in culture, and then reinstating them in the uteri of foster mothers.[14] During the period of culture, various manipulations are possible. For example, two different embryos, or cells taken from them, can be pressed together and may fuse and develop as one embryo. Information on cell fate can be obtained in this way by fusing cells from two different strains of mice or by fusing radioactively labeled cells with unlabeled cells and examining later embryos

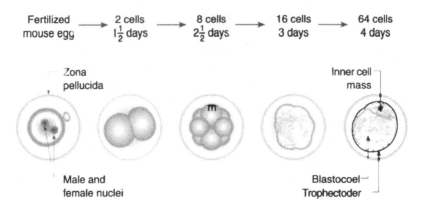

FIGURE 1-11 Early cell divisions of mouse embryo.

for cells of each type. Individual cells can also be marked by injecting certain compounds into them and their progeny cells later identified.

Cells of an eight-celled embryo have been found to have different fates: those on the outside give rise to the trophectoderm, whereas those at the center give rise the inner cell mass. Studies of 64-celled embryos have also revealed an important fact: Only part of the inner cell mass forms the mouse; the remaining cells of the embryo form the placenta and other extraembryonic tissues. That few cells of the embryo give rise to the mouse has been confirmed by fusing two 32-celled embryos with distinguishable progeny cells. Roughly 25% of the adult mice that develop from such 64-celled embryos have cells derived solely from one of the two fused embryos, suggesting that only about three cells of the embryo give rise to the adult (because if only three cells form the mouse, the chance of these all being from one of the two fused embryos is $2 \times 0.5 \times 0.5 \times 0.5 = 0.25$).

REFERENCES

1. Liebig J. *Animal Chemistry.* New York and London: Johnson Reprint Corporation, 1964.
2. Schrödinger E. *What Is Life?* Cambridge, U.K.: Cambridge University Press, 1967.
3. Perutz MF. Physics and the riddle of life. *Nature* 1987;326:555–558.
4. Driesch H. T*he Science and Philsosphy of the Organism.* London: A and C Black, 1908.
5. Mayr E. *The Growth of Biological Thought.* Cambridge, MA: Harvard University Press, 1982:118.
6. Hamburger V. *The Heritage of Experimental Embryology: Hans Spemann and the Organiser.* Oxford, U.K.: Oxford University Press, 1988:67.

7. Spemann H. *Embryonic Development and Induction.* New Haven: Yale University Press, 1938:372.
8. Gardner RL, Lawrence PA, eds. Single cell marking and cell lineage in animal development. *Philos Trans R Soc Lond B Biol Sci* 1985;312:3–187.
9. Kenyon C. The nematode *Caenorhabditis elegans.* Science 1988;240:1448–1453.
10. Sulston JE, Schierenberg E, White JG, Thomson JN. The embryonic cell lineage of the nematode *Caenorhabditis elegans. Dev Biol* 1983;100:64–119.
11. Hartenstein V, Campos-Ortega JA. Fate mapping in wild-type *Drosophila melanogaster. Roux's Arch Dev Biol* 1985;194:181–195.
12. Lawrence PA, Struhl G. Morphogens, compartments, and pattern: lessons from *Drosophila? Cell* 1996;85:951–961.
13. Slack JMW. *From Egg to Embryo: Regional Specification in Early Development,* 2nd. ed. Cambridge, U.K.: Cambridge University Press, 1991.
14. Gardner RL. Clonal analysis of early mammalian development. *Philos Trans R Soc Lond B Biol Sci* 1985;312:163–178.

FURTHER READING

Davidson EH. *Gene Activity in Early Development,* 3rd ed. Orlando, FL: Academic Press, 1986. (A detailed and scholarly analysis of fundamental experiments on many species of animals.)

Gilbert SF. *Developmental Biology,* 5th ed. Sunderland, MA: Sinauer Associates, 1997. (A superb comprehensive textbook with extensive references.)

Karp G, Berrill NJ. *Development,* 2nd ed. New York: McGraw-Hill, 1981. (A comprehensive textbook of developmental biology; old, but with basic facts exceptionally clearly described.)

Wolpert L, Beddington R, Brockes J, et al. *Principles of Development.* Oxford, U.K.: Oxford University Press, 1998. (A comprehensive textbook by a number of authoritative authors).

CHAPTER TWO

HOW AN EMBRYO ACQUIRES THE CORRECT SHAPE AS ITS CELLS DIVIDE

THE WAY A CELL DIVIDES CAN DETERMINE THE SIZE AND POSITION OF ITS DAUGHTER CELLS AND SO INFLUENCE THE SHAPE OF THE EMBRYO

A fertilized human egg cell is a round blob that can be recognized only under a high-powered microscope by an expert microscopist. Yet, in a few months this round blob can transform into a beautiful baby of complex shape with tiny hands complete with fingernails. The transformation is the result of repeated cell divisions during which controlled changes in molecular composition are introduced into the correct cells at the correct time. These changes in composition alter the behavior of the cells in a way that largely determines the shape of the developing embryo. This chapter describes how cell behavior can determine body shape—a process called **morphogenesis**.

One factor that can contribute to the shape of an animal is the position of the cleavage furrow at cell division; this determines the relative size and the positions of the daughter cells. Studies have demonstrated the importance of the cleavage furrow in determining the structure of *Caenorhabditis elegans*.[1-3] If the smaller of the two cells (P_1) formed at the first division of the fertilized egg is removed and grown in culture, it gives, as a result of the correct orientation of successive cleavage furrows, a series of cells that lie in almost the same relative positions as in the intact embryo. If the larger of the first two cells (AB) is cultured, it gives rise to an abnormal helical (corkscrew) array of progeny cells as a result of the cleavage furrows being oblique to one another. Nevertheless, this helical cleavage appears to be the basis of correct cell placement, since it is normally modified by pressure of the embryonic shell and then sustained by adhesions between cells.

Studies of the first two divisions of the fertilized egg have revealed how the resulting cells acquire their correct size and position (**Fig. 2-1**). During cell division, the contractile ring of actin filaments beneath the cell surface, which

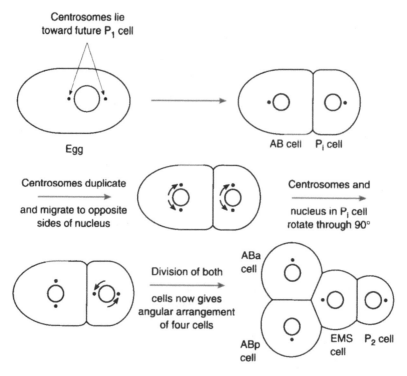

FIGURE 2-1 How the position of centrosomes determines size and position of cells in *Caenorhabditis elegans* embryo.

gives rise to the cleavage furrow, always forms halfway between the two spherical centrosomes of the spindle. The *C. elegans* egg divides to give rise to a larger AB cell and a smaller P_1 cell. This difference results from the fact that the pair of centrosomes lie toward the future P_1 cell in the dividing egg cell. Hence, the cleavage furrow also lies toward this side. After the first division, the single centrosome in each new cell divides into two, and, as usual, these each migrate 90 degrees to opposite sides of the nucleus. If both cells now divided, the cleavage furrows would form between the centrosomes to form four cells in a square. Instead, the centrosomes and nucleus in the P_1 cell rotate 90 degrees so that the following division gives the correct angular arrangement of the four new cells. Studies of mutant *C. elegans* with defects in the genes that regulate cell division are giving clues to the mechanism by which the centrosomes and cleavage furrow are positioned.

Although less important in more complex animals, orientation of the cleavage furrow often again determines the arrangement and size of cells during the first few divisions after fertilization.[4,5] In *Xenopus* and most other vertebrates, the size and position of cells in the eight-celled embryo are determined in this way. But mammalian embryos are exceptional: the first three divisions produce eight similar cells, but their placement differs from one embryo to the next. However, the cleavage furrow in the fourth set of divisions is similarly placed in every embryo and produces eight large cells on the outside and eight small cells on the inside. (The shape of a plant such as a tree is almost entirely determined in this way. The plane of each cell division in turn determines the plane along which each cellulose cell wall is laid down. Since the wall is rigid, it prevents further cell migration and movement that occurs in animals.)

The influence of the cleavage furrow on body shape is beautifully illustrated by the development of the snail *Limnaea peregra*. In the first two divisions after fertilization, the furrow has an asymmetric twist that gives an asymmetric arrangement to the daughter cells. As a result, the adult snail has a helical arrangement of its internal organs and a helical shell (**Fig. 2-2**). Since a corkscrew is asymmetric, it cannot be superimposed on its mirror image. Most adult *Limnaea* have right-handed shells that are coiled like a normal corkscrew. A few have left-handed coiling, and if these are interbred, they always give progeny with the same coiling, which can be traced back to a mirror-image orientation of the cleavage furrow in the first two divisions of the egg.

Breeding experiments show that the direction of coiling is determined solely by the embryo's mother and hence must result from the structure of the unfertilized egg. This direction is determined by a single pair of homologous genes, one derived from the mother's father and one from her mother—the gene for right-handedness being **dominant** over that for left-handedness. An asymmetric object can be produced only by an asymmetric agent: right-handed corkscrews must be manufactured by an asymmetric machine, and the

Right-handed form   Left-handed form

FIGURE 2-2 The two mirror-image forms of the snail *Limnaea peregra*.

mirror image of this machine would produce left-handed corkscrews. Presumably, the dominant gene directs the formation of a protein within the egg, which gives the cleavage furrow a right-handed twist. This protein counteracts other asymmetric molecules, which, in its absence, give a mirror-image twist. Injection before fertilization of cytoplasm from an egg of a mother with the dominant gene into an egg from a mother with two recessive genes (but not the reverse) can invert the cleavage.

CHANGE IN CELL SHAPE, LOCALIZED CELL DIVISION, AND CELL DEATH CAN INFLUENCE THE SHAPE OF THE EMBRYO

As the result of a stimulus from outside the cell, embryonic cells sometimes change **shape** and so influence the shape of the embryo. This happens in some single-layered sheets of cells (**epithelia**), causing them to fold into more complex structures. The folding of the dorsal ectoderm in frog (*Xenopus*) and newt (*Triturus*) embryos to form the neural tube, which is the precursor of the spinal cord and brain, has been studied in detail (**Fig. 2-3**). These ectodermal cells on the upper surface of the gastrula are at first cuboid, but, apparently in response to a stimulus from underlying mesoderm cells, they first elongate vertically and then contract along their upper edge. As a result, a depression forms in the cell sheet, which finally closes to form the hollow neural tube. Toward the front of the embryo, the depression is wider, and so the hollow tube has there a keyhole shape. These changes in cell shape appear to originate from within the cell rather than from outside pressure, because if the cell layer is dissected out and incubated in a culture medium, the cells elongate normally and finally fold into a tube. However, the keyhole shape arises only if underlying tissue is left attached. The folding can be reproduced by computer simulation of a sheet of cells that undergoes the same elongation and contraction. However, again, the keyhole shape results only if the program includes tension that could be caused by underlying tissue. Similar changes in cell shape appear to cause the indentation of epithelia that form the lens of the eye and to initiate the formation of the spherical sacs of salivary glands, mammary glands, pancreas, and lungs. Elongation and contraction of the cells appear to be caused by microtubules and by actin filaments, respectively, but how these are stimulated to act is unclear.

Early in the 20th century it was believed that cell division localized to a small group of cells is widespread in determining the shape of an embryo, and it was wrongly believed to cause the buckling of epithelial cells to form the neural tube. In fact, few examples have been found, but localized cell division is involved in forming the hollow sacs of mammary glands, salivary glands, pancreas, and lungs. The sacs originate from indentations that arise in epithelia

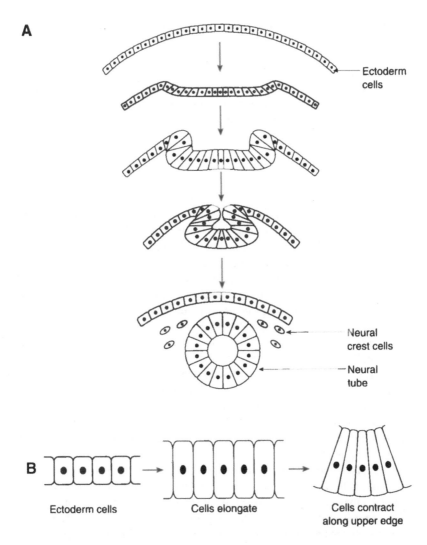

FIGURE 2-3 Formation of the neural tube in amphibian embryos (**A**) and changes in cell shape that may direct it (**B**).

by contraction of the outer surface, but the indentations are deepened to form clefts by the pressure of localized cell division. This also appears to be important in shaping the wings and legs of chick embryos and in the formation of bones.

Cell death also plays a part in determining body shape.[6,7] Sometimes, in response to hormones, it produces dramatic effects, such as the loss of tails and gills when tadpoles turn into frogs. Genetically programmed cell death,

apoptosis, is also part of the normal mechanism of development. During the development of *C. elegans*, 1090 cells are formed, but a particular 131 of these die in response to the activation of two genes (*ced-3* and *ced-4*). If either gene is inactivated, none of the 131 cells dies but without, suprisingly, any obvious ill effect. Apoptosis also occurs between the developing toes of chick and duck embryos and contributes to their shape; it also regulates the formation of immune cells in mammals where some of the genes involved have been identified.

SECRETION OF MOLECULES BY CELLS INFLUENCES EMBRYONIC STRUCTURE

A significant part of most animals consists not of cells but of extracellular molecules, mainly proteins and polysaccharides, which are secreted by **fibroblasts** and similar cells and are known as the **extracellular matrix**.[8–10] Extracellular molecules have been detected in mouse embryos with only 16 cells. As development proceeds, these molecules form an increasing part of the embryo, largely providing its framework of cartilage and bone. In adult vertebrates, over 30% of the proteins are extracellular. Extracellular molecules have two main functions: to give strength, as they do in tendon and bone, and to provide a surface to which cells can adhere, as they do in epithelia and during cell migration. The molecules that give strength are mostly **collagen** proteins. In tendons, these long collagen molecules are arranged in fibers as in a rope; those in bone are arranged in flat layers, and those in skin are arranged in a meshwork that resists stress in any direction.

Adhesion of cells to the extracellular matrix is mediated by proteins (e.g., **fibronectin** and **laminin**) known as **substrate adhesion molecules**. Some of their amino acids bind to components of the extracellular matrix such as collagen, and others bind to proteins on the surfaces of cells named **integrins**, thus linking the cell to the extracellular matrix. Integrins are composed of two parallel, cross-linked, chains of amino acids, which pierce the cell membrane. The external parts of the chains bind to one or more kinds of substrate adhesion molecules, whereas one of the internal parts binds to actin filaments within the cell. As a result, changes in external binding can affect cell shape and movement by stimulation of the cytoskeleton. Different substrate adhesion molecules and integrins have different binding specificities and hence allow vast numbers of specific interactions between cells and the extracellular matrix. The importance of integrins is shown by the inherited disease, **leukocyte adhesion deficiency**, in which one of the two amino acid chains is absent from an integrin involved in the migration of immune cells along the extracellular matrix into areas of inflammation. As a result, affected patients suffer from repeated bacterial infections.

The cells of early embryos are almost all grouped into epithelia that are stabilized by adhesion to an extracellular matrix. The matrix can also maintain the shape of folded epithelia. For example, when the matrix is removed from a developing salivary gland by enzymic digestion, the gland deteriorates into a rounded mass no longer subdivided into the sacs into which saliva is secreted. But if the enzyme is removed and the matrix is allowed to reform, the sacs reappear. The matrix also appears to play a more subtle part in development by influencing cell shape and behavior. Certain cells in the transparent tails of tadpoles can be seen under the microscope to change shape when they move from connective tissue onto the smooth wall of a blood capillary. Similar changes occur in isolated cells grown in culture dishes. That is, addition of fibronectin to many cultured cells causes them to flatten, elongate, and align with one another. Again, 3T3 cells (a **cell line** derived from mouse connective tissue) will differentiate into fat cells if cultured in suspension. However, if given a surface coated in fibronectin, they will adhere to this and flatten, but, for an unknown reason, will not differentiate. There is evidence that extracellular fibers may guide migrating cells to their correct destination.

SOME CELLS MIGRATE TO FORM TISSUES IN DISTANT PARTS OF THE EMBRYO

The importance of cell migration in shaping an embryo was not realized until well into the 20th century.[11,12] It came as a surprise when experiments showed that some tissues are formed from cells that originate in one part of the embryo and migrate to another part and that a large proportion of the body of a vertebrate is formed from such migrating cells. The earliest cell migrations in development occur in the blastula and can be easily followed under a microscope in the transparent embryo of the sea urchin (and studied in Marine Laboratories in such beautiful places as Naples or Wood's Hole.)

About 40 mesoderm cells are released into the central cavity of the blastula of the sea urchin, and they proceed to crawl individually or in small clusters over its inner surface. They eventually settle in the correct positions to form the hard spicules that protrude from the surface of the adult. Their movements resemble those of the single-celled *Amoeba*. That is, the cells put out projections that appear to explore the inner surface of the blastula, and the projections finally adhere where they make the most stable attachment. Those that adhere contract and pull the cell toward the point of adhesion.

Eggs and sperm in all vertebrates are formed from **primordial germ cells**, which have undergone a spectacular migration through the body. This has been studied in frog and chick embryos in which the migrating cells are easily identified under the microscope in stained tissue sections. In frogs, the cells become embedded in the newly formed gut at gastrulation and later migrate along the gut to the rear of the embryo, where they enter the developing sex organs. In chicks, the germ cells spend part of their journey in the developing bloodstream.

At the top of the neural tube in an early chick embryo are cells of the **neural crest** (see Fig. 2-3), which migrate to give rise to several different kinds of cell throughout the embryo. Those that migrate to form limbs have been studied in chicks. If neural crest cells are removed, a chick develops with normally shaped wings and legs but no nerve or pigment cells. The migration of cells of the neural crest can be studied by making use of quail cells. Quails and chicks are closely related, but quail cells can be distinguished under a microscope by a black spot of condensed **heterochromatin** in their nuclei. If the neural crest cells of a chick embryo are replaced by quail neural crest cells, a normal-looking chick develops, but the nerve and pigment cells in its legs and wings are found to be quail cells that must have migrated from the neural crest. These quail cells could conceivably migrate at random through the embryo and survive only in the limbs, or they could in some way be guided to their destination. Which of these possibilities is correct?

This answer has been discovered by implanting radioactive neural crest cells into the neural crest of chick embryos. Radioactive cells later become concentrated in the limbs, showing that they are somehow guided there. The paths of migration have been discovered by following the movement of quail neural crest cells grafted into chick embryos. Those that form pigment cells in the limbs migrate just below the surface, whereas those that form nerve cells migrate deeper within the embryo.

The muscles of chick wings and legs are also formed from migrating cells. Alongside the neural tube are repeating groups of cells called **somites**, which are precursors of muscle and connective tissue. If somites are cut out of a chick embryo near the place where the wings will later form, then normal-shaped wings develop, but they lack muscle cells. If the somites are replaced by quail somites, the wings contain quail rather than chick muscle cells. Examination of tissue sections shows that the quail cells migrate individually into the wings.

Migration of cells within epithelia can result in the repositioning of cells, with no alteration in their number and little alteration in their shape, and this is also important in forming tissues. The migration occurs during the conversion of the imaginal discs of *Drosophila* into adult structures, such as wings and legs, and also during gastrulation in *Xenopus*. During gastrulation, two types of repositioning occur. In one type, cells in a single-layered sheet migrate to increase their number along one axis, causing the sheet to elongate along this axis. In the other type of repositioning, cells in a two-layered sheet move between one another to form a single-layered sheet of a larger size. The initial stimulus that causes these movements is unknown.

Since migrating cells within most embryos are hidden from view, the mechanism of migration has been studied mainly in cell cultures. Connective tissue cells (fibroblasts) can be isolated from embryos by incubating them briefly with the enzyme trypsin, which weakens the adhesion between cells. Fibroblasts can then be separated by centrifugation and kept alive and growing in a glass dish on a surface of nutritive gel over which they will migrate. The cells are spherical when first placed on the gel, but after about 15 minutes, they adhere to the surface by the formation of **focal contacts** at the cell surface. They then flatten and spread. When a cell migrates, it bulges forward beyond a focal contact and then stabilizes the protrusion by forming new ones. Meanwhile, the opposite end of the cell is stretched into a tail whose focal contacts finally break under the tension, leaving small pieces of membrane and cytoplasm that adhere to the surface. The tail then contracts forward. The cell usually continues to migrate in one direction for several hours by repetition of this process. Bundles of contractile actin filaments extend within the cell from the focal contacts, and microtubules are aligned parallel with the direction of movement. These both seem to be responsible for the migration, since, if either is disrupted by chemicals that leave the cell otherwise unchanged, migration ceases.

When two fibroblasts in culture collide with one another, they stop moving. The protrusion at the leading edge of each cell retracts, and after some minutes new protrusions form on the opposite sides. The cells then usually move apart, but, if blocked by other cells, they may adhere to one another in a stable manner. This phenomenon is shown by many cells and appears to be highly important in the formation of animal tissues. A characteristic of cancer cells is that they do not usually behave in this way but continue to move in culture and pile up on top of one another. This appears to be one of the reasons why cancer cells can invade other tissues in the living body.

Most isolated cells migrate if correctly cultured, but in the developing embryo, only certain cells do so at certain times. What causes these cells to migrate in the correct direction and to stop at the correct place is unclear. Some bacteria move by **chemotaxis**; that is, they migrate from a lower to a higher concentration of a particular chemical compound. It has been suggested that embryonic cells may be directed in this way. Some white blood cells do this in culture and probably also do so in the living animal. Nerve fibers in newts appear to move toward higher concentrations of **nerve growth factor**, and their pigment cells appear to space themselves by moving away from an unidentified compound that they secrete. But there is no conclusive evidence that cells are directed by chemotaxis during embryonic development. Why migrating embryonic cells stop at the correct place is also unknown. Inhibition of further migration by contact may play a part.

There is a close relationship between the extracellular matrix and the actin filaments within cells that regulate their shape and movement. When a cultured cell secretes fibers onto a culture dish, these are often aligned parallel with the bundles of actin filaments within the cell. In turn, when other cells attach to these fibers, their actin filaments become aligned in the same direction. This suggests that extracellular fibers may guide migrating cells to their correct destination. Again, some cells draw randomly arranged collagen molecules on a culture dish into parallel fibers by movements of their actin filaments—a process that may be important in the formation of tendons and ligaments.

From experiments on cultured cells, it also appears that migrating cells may be directed along the correct paths through an embryo by gradients of increasing adhesion. For example, in one experiment, palladium metal was deposited in tiny spots on a plastic surface on which mouse cartilage cells were then cultured. The cells adhered more strongly to the metal than to the plastic. If the metal was deposited evenly, the cells migrated at random, but if deposited in a gradient of increasing closeness, the cells migrated up this gradient. Cinematography showed that the migrating cells continually form protrusions in all directions, making many adhesions to the surface. Those adhering to plastic break more easily than those adhering to palladium. Hence, the cells tend to move on to the palladium.

CELL ADHESIONS INFLUENCE THE SHAPE OF THE EMBRYO

Adhesions between cells do not appear to merely stabilize tissues after their formation but may direct the assembly of cells into tissues. This became clear about 40 years ago, when it was shown that similar cells adhere to one another in preference to dissimilar cells and that they may spontaneously assemble into tissues. For example, in one experiment, liver and cartilage were dissected from chick embryos and dissociated into single cells by brief treatment with the enzyme trypsin. (Trypsin digests proteins and hence molecules involved in cell adhesion, but most of these are re-formed after a few hours.) The two kinds of cell were mixed in a culture medium, and they adhered into clusters that contained both kinds of cells. However, over about 48 hours, the two kinds of cells separated from one another or "sorted out" into clusters of a single kind and even assembled into some of the structures from which they came, such as the bile ducts of liver.

It was suggested that sorting out is founded on different strengths of adhesion and resembles the separation of two immiscible (incapable of mixing) liquids whose molecules adhere with different strengths. If two such liquids are shaken and the emulsion left to stand, the molecules of greater adhesiveness form droplets of gradually increasing size within the liquid of lower adhesiveness. Similarly, when heart and retina cells from a chick embryo were mixed, the heart cells formed clusters of gradually increasing size within a rounded mass of retina cells. The ways in which various cell mixtures sorted out supported the theory that the sorting out depends on different strengths of adhesiveness between each type of cell. For example, in one experiment three kinds of chick embryo cells were prepared: from limb bud, heart, and liver. The three possible mixtures of two kinds of these cells were made and allowed to sort out. In the heart + limb bud mixture, heart cells came to surround limb bud cells, whereas in the heart + liver cell mixture, liver cells came to surround heart cells. This suggests that the strength of adhesion between similar cells is limb bud > heart > liver. If so, a limb bud + liver mixture should sort out with liver cells on the outside, and it did so.

Sorting out by this mechanism could result from all cells having adhesion sites of one kind on their surface, which adhere to identical sites on another cell, but different kinds of cells having different numbers of sites per unit area of membrane. However, other experiments suggested that adhesion sites can be of more than one kind, adhering only to similar sites on another cell. For example, in one experiment, cells were isolated from the retina and liver of chick embryos, some of the cells having been radioactively labeled. The unlabeled cells were shaken until they adhered into small aggregates of liver or retina cells. Then, one of the two types of aggregates was shaken with one of the two types of labeled cells in four different combinations. In a series of experiments, an average of 23 retina cells, but less than one liver cell, adhered to retina cell aggregates, whereas an average of 10 liver cells, but only one retina cell, adhered to liver cell aggregates. It is clear that the adhesion sites on liver cells must differ from those on retina cells

These experiments suggested that cell adhesions are important in the assembly of cells into tissues, and experiments on living embryos confirmed this. For example, a newt embryo at the early gastrula stage is composed of three distinct cell layers: endoderm, mesoderm, and ectoderm, which later form different body tissues. Newt embryos were radioactively labeled by growing them in a radioactive medium, and cells were isolated from these three cell layers. Each type of cell was injected into unlabeled embryos at the earlier blastula stage when the three types of cells have not yet formed. When these embryos later formed gastrulae, labeled endoderm cells were found in endoderm, labeled mesoderm cells in mesoderm, and labeled ectoderm cells in ectoderm.

TYPES OF CELL ADHESIONS

Adhesions arise both from stable **cell junctions**, which occupy a limited part of the cell surface and have well-defined structures revealed by electron microscopy and from more mobile interactions between individual **cell adhesion molecules**. They are also reinforced in less specific ways. These include weak attractions that exist between all molecules, weak bonding between sugar molecules that protrude from cell membranes, and bonding of protein molecules on one cell membrane to sugar molecules on another.

Cell junctions form between cultured cells, which enables their structure to be studied. They can be subdivided into **tight junctions**, **anchoring junctions**, and **gap junctions**. Tight junctions are composed of protein strands that encircle the upper ends of epithelial cells and bind them closely together, thus helping to maintain the structure of the epithelium. During development, these junctions are first found in the blastula. Anchoring junctions occur in many cells and are composed of proteins that pass through the cell membrane. Inside the cell, they bind to microfilaments and outside to the extracellular matrix, or to proteins similar to themselves protruding from neighboring cells. The main function of anchoring junctions is to stabilize tissue structure, and they are first found in the late blastula. Gap junctions are composed of protein molecules grouped into hollow cylinders that pierce the cell membrane. Cylinders from two adjacent cells are joined and form a channel that allows small molecules to pass between the cells. This is proved by injecting fluorescent molecules into a cell: they pass into adjacent cells only if they are joined by gap junctions, and injection of antibodies against gap junction proteins inhibits the transfer. As a result, gap junctions can enable synchronous action of adjacent cells, such as heart muscle cells, by allowing small regulating molecules to have the same concentration in all cells. But they also contribute significantly to the stabilization of tissues, and they occur early in an animal's development. If antibodies against gap junctions are injected into one cell of an eight-celled *Xenopus* embryo, its development is disrupted.

Cell adhesion molecules were discovered more recently in experiments purposely designed to search for them. In one experiment, retina cells from chick embryos were injected into rabbits to evoke the formation of antibodies to molecules on the cell surface. Antibody molecules were isolated from the rabbit's blood and found to prevent the adhesion of retina cells to one another, presumably because some antibodies bound to cell adhesion molecules on the cell surface and so inactivated them. A complex sequence of protein fractionations finally gave a **pure** (i.e., all molecules identical) sample of a cell adhesion molecule. It was named **neural cell adhesion molecule** (abbreviated to N-CAM). It was called "neural" because retinal cells are nerve cells. When

N-CAM was incorporated onto the surface of lipid droplets, these bound to one another and to retina cells, and the binding was inhibited by antibodies to N-CAM. The molecule is a single protein chain of about 1000 amino acids, with sugar molecules attached to some of these amino acids. Other adhesion molecules were identified on other cells. One is the **liver cell adhesion molecule** (L-CAM). Although discovered in the liver, L-CAM occurs in all epithelial cells of embryos and adults.

Cell adhesion molecules are proteins that protrude through the cell membrane. The protruding portion can bind to the same portion of another (usually identical) molecule on another cell. Most of the molecules fall into two groups. Those now called **cadherins** adhere to another protein of identical structure on another cell, but only if calcium ions are present (which explains why a low calcium concentration often aids the isolation of cells from tissues). The end of the molecule within the cell is attached to the cytoskeleton by other proteins called **catenins**. As a result, changes in external folding of the protein chain may be transmitted to the cytoskeleton and initiate molecular changes within the cell. Over 30 cadherins have been identified, including L-CAM. Those of the second group form the **immunoglobulin superfamily** of adhesion molecules. The structure of this group of molecules is similar to that of antibody proteins, which are thought to have originated from them. In addition, other cell surface proteins adhere to molecules of the extracellular matrix, such as collagen, fibronectin, and laminin. The most clearly identified are the **integrins**. They are composed of two adjacent protein chains each of which can vary in structure so giving rise to many integrins of which over 20 are known. Within the cell they are again associated with the cytoskeleton.

THE IMPORTANCE OF CELL ADHESION MOLECULES

The study of cell adhesions has become one of the main activities of research into animal development, because cell adhesions may be the principal agents of **morphoregulation**, the arranging of cells into the correct body structure.[13,14] Study has been stimulated by the fertile ideas of Gerald Edelman, who was awarded the Nobel Prize for his part in discovering the structure of antibody proteins and who later, with his colleagues, isolated the first cell adhesion molecule. A number of experiments have demonstrated the importance of these molecules in the adhesion of cells and suggest that they are one of the main driving forces of animal development.

The distribution of adhesion molecules can be studied by staining microscopic sections of embryos with fluorescent antibodies to these molecules and examining the sections in a fluorescence microscope. The different kinds of adhesion molecules are found to be widely distributed, and each kind of tissue tends to have characteristic adhesion molecules. As new tissues are formed in an embryo, local changes in adhesion molecules occur, which are consistent with their part in determining tissue structure. For instance, during compaction of the eight cells of a mouse embryo (see Fig. 1-11), E-cadherin (i.e., L-CAM) becomes concentrated in the region of cell adhesion and is converted from an inactive to active form. Again, during the development of organs, cells express different adhesion molecules as they separate into different regions, as shown during the formation of feathers in chicks and of the neural tube in *Xenopus*. Moreover, when nerves carrying N-CAM approach a muscle to which they will connect, N-CAM appears on the muscle surface; antibodies against N-CAM disrupt these connections.

The binding of adhesion molecules on different cells is usually followed rapidly by the formation of cell junctions. Adhesion of cells often changes their shape from flat to round or vice versa, probably through the attachment of adhesion molecules to the cytoskeleton. This can cause changes in cell composition, such as pH, and evidence suggests that these changes may activate cell division and gene activity and so contribute further to the formation of body structure.

Migrating cells can be studied with an electron microscope in thin sections of embryos. Their membranes are often very closely aligned to those of the cells over which they are moving, which is consistent with their adhering by cell adhesion molecules. It is suggested that the distribution of each kind of adhesion molecule varies within the embryo and changes with time in a way that directs cells along the correct paths of migration. The distribution of these molecules in mouse, chick, and *Drosophila* embryos has been studied by staining sections of the embryos with antibodies to the molecules. The antibodies have a fluorescent dye attached, and where they bind within the embryos, they can be seen in a fluorescence microscope. The different kinds of cell adhesion molecule are found to be widely distributed, and as development proceeds they undergo local changes in concentration consistent with their directing migration.

An elegant experiment has shown that adhesion molecules could be important in the sorting out of dissimilar cells.[15] Mouse **L-cells** (an epithelial cell line) have been made to express cadherins on their cell surface by injection of genes that direct cadherin formation into their cell nuclei. The cells, which normally do not adhere, now do so to form an epithelium, provided that they are in a medium containing calcium ions. Two strains of L-cell were produced, the first having 20 times as much P-cadherin on its surface as the second. In a medium containing calcium ions, cells of the first strain aggregated rapidly, and those of the second strain aggregated slowly. Pellets composed of both kinds of cell were left on agarose gel for 4 days in a closed dish. The cells sorted out to give a central sphere of cells of the first kind, surrounded by a shell of cells of the second kind. Also, when separate pellets of the two kinds of cell were left in contact, cells of the second kind migrated over those of the first to give a similar structure. This experiment shows that cells with different numbers of the same adhesion molecule on their surface will sort out from one another. This suggests that different kinds of adhesion molecules are not always required for different cells to assemble into different tissues.

Adhesion molecules of a different kind, **bindins**, have been discovered on some sperm, which make them adhere only to eggs of the same species. Bindins are found in many aquatic animals whose sperms are shed into the water, where they mingle with the eggs of many other species. These molecules have been studied in sea urchins. For example, in one experiment eggs and sperm from three species of sea urchin were collected, and all nine possible mixtures of eggs and sperm were made. When the eggs and sperm were from the same species, over 96% of the sperm bound to the eggs. But when they were from different species, less than 5% of the sperm bound to the eggs. Two proteins involved in this adhesion have been isolated—one from sperm (the bindin) and another from eggs (the bindin receptor). As expected, the sperm protein binds only to eggs of the same species, and if the egg protein has been first added to the sperm protein, it inhibits this binding.

REFERENCES

1. Hyman AA. Centrosome movement in the early divisions of *Caenorhabditis elegans*. *J Cell Biol* 1989;109:1185–1193.
2. Strome S. Determination of cleavage planes. *Cell* 1993;72:3–6.
3. Horvitz HR, Herskowitz I. Mechanisms of asymmetric cell division. *Cell* 1992;68: 237–255.
4. Freeman G, Lundelius JW. The developmental genetics of dextrality and sinistrality in the gastropod *Lymnaea peregra*. *Wilhelm Roux's Arch* 1982;191:69–83.
5. Tamura K, Yonei-Tamura S, Belmonte JCI. Molecular basis of left-right asymmetry. *Dev Growth Differ* 1999;41:645–656.
6. Metcalfe A, Streuli C. Epithelial apoptosis. *BioEssays* 1997;19:711–720.
7. Vaux DL, Korsmeyer SJ. Cell death in development. *Cell* 1999;96:245–254.
8. Watt FM. The extracellular matrix and cell shape. *Trends Biochem Sci* 1986;11:482–485.
9. Adams JC, Watt FM. Regulation of development and differentiation by the extracellular matrix. *Development* 1993;17:1183–1198.
10. Hynes RO. Genetic analyses of cell-matrix interactions in development. *Curr Opin Genet Dev* 1994;4:569–574.
11. Hynes RO, Lander AD. Contact and adhesive specificities in the associations, migrations, and targeting of cells and axons. *Cell* 1992;68:303–332.
12. Howard K, Jaglarz M, Zhang N, et al. Migration of *Drosophila* germ cells: analysis using enhancer trap lines. *Development* 1993;(Suppl):213–218.
13. Edelman GM. Morphoregulation. *Dev Dyn* 1992;193:2–10.

14. Huttenlocher A, Sandborg RR, Horwitz AF. Adhesion in cell migration. *Curr Opin Cell Biol* 1995;7:697–706.
15. Steinberg MS, Takeichi M. Experimental specification of cell sorting, tissue spreading, and specific spatial patterning by quantitative differences in cadherin expression. *Proc Natl Acad Sci USA* 1994;91:206–209.

FURTHER READING

Bard J. *Morphogenesis: The Cellular and Molecular Processes of Developmental Anatomy.* Cambridge, U.K.: University Press, 1990.
Hogan BLM. *Morphogenesis.* (Review.) *Cell* 1999;96:225–233.
Trinkaus JP. *Cells into Organs: The Forces That Shape the Embryo,* 2nd ed. Englewood Cliffs, NJ: Prentice-Hall, 1984. (The background to current work is particularly well discussed and cited.)

CHAPTER THREE

How Differences in Molecular Composition between Embryonic Cells Originate

Chapter 1 described the different ways in which cells divide during animal development, each being determined by the particular molecular composition of the parent cell. Chapter 2 described how the shape of an embryo is largely determined by the properties of its component cells; these are again determined by their molecular composition. Hence, the fundamental problem of development is how, as cells increase in number, each cell acquires the correct molecular composition at the correct time. The remainder of the book is devoted to this problem.

ASYMMETRIC CELL DIVISION AND SIGNALING BETWEEN CELLS

An adult human has over 200 types of cells that are visibly different and hence differ in molecular composition. Such differences arise progressively during the repeated cell divisions that convert the fertilized egg into the adult. Experiments have shown that these differences arise in one of two ways: by asymmetric division of a parent cell to give rise to two cells of different composition or, without division, in response to a stimulus or "signal" received by a cell and normally emitted by another cell. Such signals are usually molecules that contact the cell and give rise to a sequence of molecular interactions, or **signal transductions**, which result in the required change in composition. Each kind of animal was once thought to develop largely by a particular one of these two mechanisms, which were named mosaic and regulative—*mosaic* because repeated asymmetric divisions could give rise to a mosaic of cells whose initial differences could be amplified by interaction of their component molecules and *regulative* because one cell is regulated by molecules emitted by another.

It is now clear that although in the first few divisions after fertilization differences in composition usually arise from asymmetric division, in most animals signals soon become more important and are transmitted between different kinds of cells that become temporarily aligned by the complex movements of cell layers that begin during gastrulation. In the early development of many insects such as *Drosophila*, in which nuclei are not separated by cell membranes, these two mechanisms are modified, but remain essentially similar. Either daughter nuclei are placed in cytoplasm of differing composition, as occurs after asymmetric cell division and, as a result, are differently activated, or one nucleus is activated by a signal derived from another.

Around 1900, the famous German embryologist, Hans Spemann,[1] was among the first to show that differences in composition that determine the fate of cells can arise by either asymmetric division or signaling. He studied the effects of separating the cells of early newt embryos by tightening a loop of the fine hair from his baby son between them. The egg of a newt (*Triturus*) is very similar to a *Xenopus* egg, and it usually divides after fertilization in a similar way (see Fig. 1-9). Occasionally, the first division occurs along a different plane but still gives rise to a normal embryo: it gives rise to one cell that contains the point of sperm entry and another that contains the opposite pole of the egg, which in newts (but not in *Xenopus*) is marked by a **gray crescent**. When these two cells were separated, the cell with the gray crescent usually developed into a normal embryo, but the other merely formed an abortive ovoid structure, which sometimes contained kidney and blood cells. Spemann drew two conclusions from this result.

First, because the cells developed differently, the components of the fertilized egg must have been unevenly distributed between the daughter cells by an asymmetric division. Second, since one cell developed abnormally, this cell or its progeny must require signals from the other cell or its progeny for correct development. In Spemann's words, the signal might be "an unorganized substance needed for the release or formation of the organs, or organized material that can differentiate into these organs and perhaps also incite other cells to differentiate."[1]

Around 1930, the importance of asymmetric cell division in early development was clearly demonstrated by performing experiments on sea urchins by Sven Hörstadius.[2–4] The development of the many closely related species of sea urchins has been studied for nearly 200 years. Their eggs, being only about one-tenth of a millimeter (0.1 mm) in diameter, are only just visible to the naked eye. Under the microscope, the animal and vegetal poles can be distinguished in structure, implying differences in composition within the egg. After fertilization, four successive and synchronous cleavage divisions subdivide the egg into 2^4 (16) cells, and further cell divisions give rise to the larva (**Fig. 3-1**). In the first two divisions, the planes that separate the two new cells pass vertically through the animal and vegetal poles, with the second plane at right angles to the first. The plane of the third division is horizontal through the equator, giving rise to four

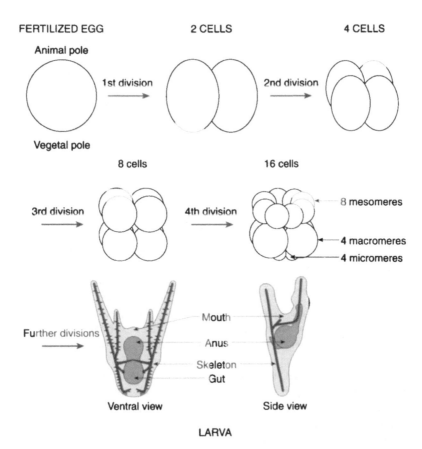

FERTILIZED EGG   2 CELLS   4 CELLS

Animal pole

1st division   2nd division

Vegetal pole

8 cells   16 cells

3rd division   4th division   8 mesomeres

4 macromeres

4 micromeres

Further divisions

Mouth
Anus
Skeleton
Gut

Ventral view   Side view

LARVA

FIGURE 3-1 Cell divisions of the sea urchin embryo.

cells at the animal pole and four at the vegetal pole, all roughly of equal size. The next divisions produce a ring of eight cells at the animal pole (**mesomeres**), a ring of four larger cells beneath them (**macromeres**), and another ring of four smaller cells at the vegetal pole (**micromeres**).

The positions of these cells are identical in every embryo, and each cell gives rise to a characteristic group of progeny cells. This suggests that these three rings of cells are directed along different courses of development by molecules that are unevenly distributed in the fertilized egg and are separated into the different cells by the cleavage divisions. Simple experiments confirmed this. They involved separation of groups of cells from embryos and their recombination in various arrangements. Immediately after fertilization, the gel coat and the hard outer membrane were removed from the eggs and, after cell division, groups of cells were teased apart with a needle in calcium-free sea water to weaken their adhesion. They were reassociated by dropping them into ordinary sea water in tiny cylindrical depressions in a plastic plate and pressing them together with a glass rod.

When the four cells resulting from the first two divisions of the egg were separated, each developed into a small, but more or less normal, larva. Evidently, the egg is radially symmetric about a line from the animal to the vegetal pole, thus giving each cell the same content of essential molecules. These cell molecules in turn adjust back to the original distribution within the egg. However, when eight-celled embryos were separated into the two groups of four at the animal and vegetal poles, they developed differently. The animal halves merely formed hollow spheres of ciliated cells, whereas the vegetal halves formed larvae with normal internal skeleton, but with other body structures distorted. These larvae were each similar to the abortive embryos formed when an unfertilized egg was divided horizontally and the two halves fertilized separately. It was concluded that the animal and vegetal halves of the egg contain different molecules, which are separated by the third division and are responsible for initiating development along different paths. However, the fact that each half of these embryos developed abnormally suggested that normal development also requires an interchange of signals between the two halves of an embryo. This was confirmed in further experiments.

As already mentioned, the 16-celled embryo of the sea urchin has three tiers of cells. When the tier of four cells at the bottom of the embryo was separated and cultured, the cells went on to divide the normal number of times and formed their normal structures, namely, the **skeletal spicules**. These cells and their progeny cells appeared to require no signals from other parts of the embryo for normal development. However, when the tier of eight cells at the top was separated from the remainder, the two halves developed into the hollow sphere and the distorted embryo already described. But if the two halves were immediately reassociated, development was normal, showing that signals must pass between them. The four cells at the bottom of the embryo appeared to be the major source of the signals required by other cells: when these four cells were implanted beneath the 16 cells taken from the upper half of a 32-celled embryo, an almost normal sea urchin developed. The four cells gave rise to skeletal cells as usual and induced cells above them to form gut and other tissues that were lacking when the upper half developed alone.

ASYMMETRIC DIVISION AND CELL SIGNALING IN *C. ELEGANS*

As seen in the last section, most fragments of sea urchin embryos develop abnormally, suggesting their need for signals from the remainder of the embryo. However, in many simple organisms, including nematodes such as *Caenorhabditis elegans* (see Chapter 2), the development of groups of cells is often unaltered by their excision from the embryo and growth in culture. It was suggested that development of these organisms is entirely mosaic. However, studies over recent years have shown that cell differences in *C. elegans* arise by both mosaic and regulative mechanisms.

Experiments have shown very clearly that the *C. elegans* embryo has a mechanism for segregating certain molecules to a particular one of two daughter cells at division, as would be required for mosaic development.[5,6] They result from observations of **P-granules**, which are small particles visible in the electron microscope. In the adult, P-granules are confined to germ cells, which divide to give eggs and sperm. Immediately after fertilization of the egg, the P-granules are evenly dispersed but soon drawn by actin filaments to one side of the egg. As a result, after the first division, they are confined to the outer side of the P_1 cell. The subsequent migration of the centrosomes in this cell (see Chapter 2) ensures that the P_1 cell divides so that the granules are confined to the outer cell that is formed. At each of the next two divisions, the granules segregate into the daughter cell whose progeny includes germ cells, and at subsequent divisions they are evenly distributed between daughter cells, all of which later form germ cells.

There is no evidence that P-granules are involved in the formation of germ cells. However, molecules that cause changes in cell composition could well be similarly segregated, and experiments on the formation of an enzyme peculiar to gut cells suggest that they are.[7] The enzyme (an esterase) is easily revealed in cells by a staining reaction. It can first be detected in embryos with over 100 cells and is confined to a few cells whose progeny later form gut cells. Its formation has been shown to result from activation of the gene that directs the formation of the enzyme. Although embryos with less than 100 cells do not contain the enzyme, the cells whose progeny form gut cells do appear to contain molecules that lead to activation of the gene. This has been demonstrated by stopping cell division—but not stopping all changes in molecular composition—by adding the drug, cytochalasin. When division is stopped in the two-celled embryo, the enzyme is formed after further incubation in the now enlarged rear cell. When it is stopped in embryos with 4, 8, 16, or more cells, the enzyme is again formed on further incubation only in cells that have gut cells as progeny.

Similar experiments on other simple organisms leave little doubt that the differentiation of certain cells requires molecules that are transmitted to these cells from the fertilized egg by a sequence of asymmetric divisions. Also, when single cells of embryos were destroyed with a laser beam, they often gave rise to adults that lacked only tissues formed from the destroyed cell. As a result, it was generally agreed that such organisms largely undergo mosaic development with little need for signaling between cells. However, over recent years evidence has accumulated that signaling between the cells of *C. elegans* embryos is extensive.[8] For example, the vulva is formed late in development by the division of three cells. If these are destroyed by a laser beam, neighboring cells move into their place to form the vulva. These cells must receive a signal in their new position, which causes them to change their fate, and its source proves to be an overlying **anchor cell**. When this cell only is destroyed, the three cells that normally form the vulva do not do so but develop like neighboring cells (see also Chapter 7).

In other experiments on *C. elegans*, particular cells of the early embryo were removed or rearranged with a micromanipulator. On division of the fertilized egg of *C. elegans*, the anterior cell divides further to give rise two cells, the anterior one of which gives rise to muscle cells of the pharynx. However, if the positions of these two cells are interchanged, the other cell now gives rise to the muscle cells of the pharynx. The anterior cell is clearly receiving a signal, and its source has been identified: when the posterior cell of the two-celled embryo (or the anterior one of its two daughter cells) is removed, the muscles are not formed. Several other signals have been shown to be essential for development of the fertilized egg into an embryo with 50 cells. Work on the development of *C. elegans* is progressing rapidly and is revealing the molecular details of both its mosaic development and its signaling (see Chapter 7).

UNEVEN SUBDIVISION OF THE EGG CYTOPLASM IN *DROSOPHILA*

The *Drosophila* (fruit fly) egg nucleus first subdivides without cell division to give rise to several thousand nuclei. Most of these nuclei migrate to form a single layer below the surface of the egg. Only later do membranes surround these nuclei to form cells, and until then there is no possibility of asymmetric cell division or signaling between cells. However, early experiments showed that the egg cytoplasm is not uniform in composition and that nuclei are induced to later form different cells by coming to lie in cytoplasm of different composition. The clearest evidence came from studies of pole cells—the first cells to be formed in the embryo.[9] These pole cells arise from about 15 nuclei that do not migrate to the surface of the egg but gather at its rear end and later give rise to the germ cells (egg or sperm cells; see Fig. 1-5). Shining a beam of ultraviolet light on this region of the egg before the nuclei moved there was found to disrupt the formation of pole cells and give rise to sterile adults. The implication was that the posterior cytoplasm contains molecules—destroyed by ultraviolet light—which initiate the formation of germ cells.

This conclusion was confirmed by further experiments in which nuclei that do not normally form germ cells were made to do so by surrounding them with posterior cytoplasm (**Fig. 3-2**). For example, 5 microliters (5 μL) of posterior cytoplasm (without nuclei) was withdrawn from a number of normal (wild-type) *Drosophila* embryos. Each sample was injected into the front of an embryo of a mutant strain (strain 1) whose adults (because they contained two copies of a recessive mutant gene) could be identified by their characteristic appearance. Nuclei in the front of these embryos were now found to become

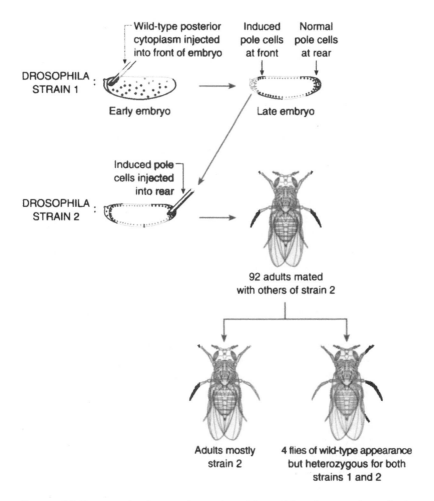

FIGURE 3-2 Demonstration that cytoplasm at rear of *Drosophila* embryo contains molecules that activate the formation of germ cells.

incorporated into cells with the distinctive structure of pole cells. Would these cells now give rise to germ cells of strain 1?

To test whether such germ cells would be produced, the pole cells were removed with a glass needle and injected among the pole cells at the rear of embryos of a second recessive mutant strain (strain 2), whose adults had a different appearance. From 228 of such embryos, 125 adult flies developed, of which 92 were fertile (damage during the injections made development less successful than normal). Were some of the egg and sperm cells of these fertile strain 2 flies derived from injected cells of strain 1, which had previously acquired the structure of pole cells by being immersed in posterior cytoplasm? To test this, all 92 fertile flies were crossed with others of strain 2. Most of the resulting progeny resembled this strain, showing that they had originated from the union of an egg and a sperm that both carried the recessive gene of strain 2. However, four of the flies resembled those produced by crossing strain 1 with strain 2; that is, they resembled normal wild-type flies. Breeding experiments showed that, as expected, this was because they were heterozygous and carried only one copy of each mutant gene. Hence, these four flies must have arisen from a sperm of strain 1 fertilizing an egg of strain 2, or vice versa. Therefore, some of the nuclei at the front of embryos of strain 1 must have been directed to form germ cells by the posterior cytoplasm with which they were artificially surrounded.

Other experiments suggested that nuclei that migrate to beneath the surface of the *Drosophila* embryo are also directed to different fates by different molecules in different regions of the cytoplasm. Thus, when cells that had recently been formed by enclosure of these nuclei in cell membranes were isolated and implanted into other *Drosophila* embryos, they gave rise to identifiable adult structures. Cells from the anterior of early embryos gave rise to only anterior structures, such as parts of eyes, legs, and wings, whereas those from the posterior gave parts of the thorax and abdomen at the rear of the adult body.

The latter experiments led to the suggestion that all differences in gene activity between nuclei in the early *Drosophila* embryo are induced by static regional differences in the composition of the egg cytoplasm. But studies of egg formation within female *Drosophila* made it difficult to explain how the required complex mosaic of inducing molecules could be laid down. Instead, experiments on a number of insects, including *Drosophila*, suggested a different mechanism.[10] Embryos were temporarily constricted— either by ligature with a thread or by pressure of a razor blade—to reduce communication between cytoplasm on each side of the constriction. If early embryos contained a static mosaic of inducing molecules, then development should not have been affected by constriction (provided that it did not damage the embryo, and comprehensive tests suggested that it did not). When later embryos were constricted to separate the front and rear halves, they did develop normally. But when this was done earlier, the embryos developed with a series of healthy cells from front to rear; however, these were incorrectly programmed in that they did not give rise to the correct body segments.

Many insects and other arthropods have segments or externally repeating units along their length, as is well known to children who have played with centipedes. Both *Drosophila* larvae and adults have 14 segments (**Fig. 3-3**), and they can be distinguished from one another in the larva by each having a characteristic arrangement of bristles. In embryos that had been constricted, segments that should appear around the center of the larva were absent. For

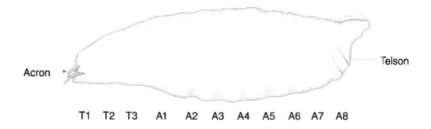

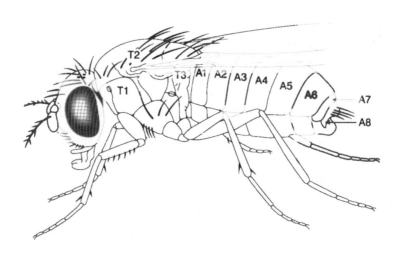

FIGURE 3-3 Thoracic (T) and abdominal (A) segments in the larva and adult *Drosophila*. The acron and telson are unsegmented structures at the front and rear.

example, the epidermal cells might form segments 1 through 6 and then abruptly change to forming segments 13 and 14.

It was concluded that nuclei at the center of the embryo acquire the correct pattern of gene activity only if cytoplasm at the front can communicate with cytoplasm at the rear. It was suggested that two different kinds of molecules are released from the front and rear of the embryo and that each diffuses toward the opposite end to form a gradient of falling concentration. The local ratio of concentrations along the embryo would induce in each nucleus a pattern of gene activity appropriate for its position.

When insect embryos were constricted not across their length but along it, so that communication between the two sides was hindered, the effects were different. That is, two normal embryos often formed, joined side by side. The important conclusion was drawn that two separate inducing gradients must act—one along the anterior-posterior axis (front to rear) and the other at right angles to it along the dorsal-ventral axis (above to below). Hence, each nucleus would acquire the pattern of gene activity appropriate for its position along these two axes and so give a series of cells that varied in structure in both directions. These experiments, which were performed over 20 years ago, appear naively simple, but the conclusions hit several nails on the head. This will be seen when recent complex molecular and genetic experiments on *Drosophila* are described in Chapter 6.

ASYMMETRIC DIVISION AND CELL SIGNALING IN FROGS

Different regions within the egg of the frog *Xenopus* are assigned to different cells by asymmetric cleavage divisions (see Fig. 1-9). Studies of **germ plasm**, a region of cytoplasm of distinctive appearance near the vegetal pole, suggest that in this way the cells can acquire molecules that determine their different fates. This cytoplasm is unevenly distributed among cells during early divisions, and in a 32-celled embryo, it is confined to a few large pole cells. The fate of these cells can be followed by their peculiar appearance. They later migrate to the developing sex organs and form around 30 primordial germ cells, which give rise to eggs or sperm. If the germ plasm is removed from an egg through a fine glass needle or damaged by local irradiation with ultraviolet light, the number of germ cells is reduced. This effect can be reversed by injecting germ plasm from another egg. If extra germ plasm is injected into a normal egg, more germ cells are formed. These facts suggest that certain molecules localized at the vegetal pole of the egg direct cells that enclose them toward becoming germ cells.

Other experiments[11] have shown that differences in composition between cells formed in the first two cleavage divisions cause them to follow different paths of development. The planes of these divisions both pass from the animal to the vegetal pole—the first along the dorsal-ventral axis to give the right and left sides of the embryo and the second in a plane at right angles to this to give the upper and lower halves. Experimentally. eggs were fertilized, and the gel and protective vitelline membrane surrounding the egg were removed. After division, the cells were drawn apart with a hair loop, excised with a glass needle, and left to develop on their own. Cells from about 100 two-celled embryos were first separated and left to develop further. Both cells from about one-half of the embryos, and one cell from another fourth, developed into small but almost normal tadpoles that usually grew into frogs.

Such predominantly normal development demonstrated that the fertilized egg first divides into two halves of essentially identical composition. However, when four-celled embryos were divided into pairs of cells along a plane at right angles to the previous one (i.e., into dorsal and ventral halves), neither pair developed normally. A few of the ventral halves gave embryos lacking heads, with the rest of the body normal, although most gave abnormal masses of cells. The dorsal halves usually gave rise to embryos with enlarged heads and small tails. It was concluded that molecules that direct cells along the correct paths of development are unevenly distributed in the egg and unevenly distributed between cells formed at the second division. Molecules that are unevenly distributed in the egg and could well be involved have recently been discovered, namely, messenger RNAs for protein growth factors, as explained in later chapters.

However, experiments have shown that further differences between cells soon arise in response to signaling, namely, in the formation of mesoderm cells.[12] Inspired by the experiments of Hörstadius[2] on sea urchins, blastulae of *Xenopus* were cut into four transverse sections between the animal pole (section 1) and vegetal pole (section 4). When sections 1 or 2 were incubated for 4 days, they formed ectoderm cells (e.g., nerve), whereas section 4 formed endoderm cells (e.g., gut). The different kinds of cell were presumably induced by different molecules inherited from different regions of the egg cytoplasm. However, when sections 1, 2, and 4 were reassociated in the correct order, about 20% of the cells that formed were mesoderm cells (e.g., muscle). It was concluded that these cells arise from an exchange of signals between cells of the animal and vegetal regions. The question then arose: Are mesoderm cells formed from animal cells that would otherwise give rise to ectoderm cells, from vegetal cells that would otherwise give rise to endoderm cells, or from both? When sections 1 and 2 from radioactive blastulae were reassociated with section 4 from a normal blastula, the mesoderm cells were radioactive, showing that they are formed from animal cells in response to a signal from vegetal cells.

Further experiments showed that vegetal cells emit more than one kind of signal (**Fig. 3-4**) and induce more than one kind of mesoderm cell. Section 4 was subdivided to compare the ability of different regions to induce mesoderm cells in sections 1 and 2. Dorsal fragments of section 4 (from opposite the sperm entry point) were the most powerful and induced a different type of mesoderm, including more muscle and fewer blood cells. By testing single cells from 32-celled blastulae of *Xenopus*, this inducing ability was found to be confined to two cells, namely, C1 and D1 (see Fig. 1-9). These are now called the **Nieuwkoop Center** after the originator of the experiments. The activity of the vegetal cells results in some way from rotation of the cortical cytoplasm soon after fertilization. This probably stimulates the formation of signaling molecules from precursors unevenly distributed in the egg. If rotation is inhibited by x-radiation to disrupt microtubules, the characteristic mesoderm cells are not induced; if rotation is artificially restored, the cells are induced.

The dorsal mesoderm cells that are induced by the Nieuwkoop Center in turn emit another signal that induces further differences in composition toward the ventral side in adjacent mesoderm cells. This was shown by separating the newly formed mesoderm from each side of the embryo.

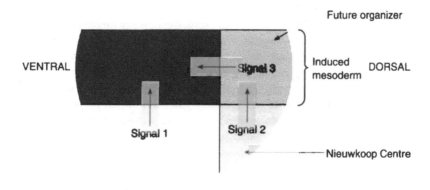

FIGURE 3-4 The three-signal model of mesoderm induction in the early embryo of *Xenopus*. Signals 1 and 2 from the vegetal side of the embryo induce different kinds of mesoderm cells at the equator. Signal 3, released by mesoderm on the dorsal side, induces further changes toward the ventral side.

When cultured alone, mesoderm from the side of sperm entry did not form muscle cells, but when associated with mesoderm from the opposite dorsal side, it did so.

In summary, it was concluded that three kinds of mesoderm inducing signals are active (see Fig. 3-4).[13,14] The first kind of signal is released by most cells of the vegetal region and transforms the cells above them into one kind of mesoderm. However, two of the vegetal cells at the dorsal side release a different signal, which induces mesoderm cells of another kind. These dorsal mesoderm cells then release a third signal that moves toward the ventral side and modifies the type of mesoderm cells formed by the first signal. As discussed in Chapter 4, the signals that induce mesoderm cells are the first in a complex series of inducing signals that continue throughout the development of vertebrates.

THE ORIGIN OF DIFFERENCES IN CELL COMPOSITION IN MOUSE EMBRYOS

Mouse embryos, like other mammalian embryos, are unusual in a number of ways. When the embryo has 64 cells, about 48 form the trophectoderm, a spherical outer layer one cell thick, which later forms extraembryonic tissues (see Fig. 1-11). Within this sphere lies the inner cell mass, only part of which gives rise to the mouse, the remainder again giving rise to extraembryonic tissues. The fertilized mouse egg, unlike that of the frog, does not have regions of different composition that are subdivided among daughter cells to direct them along different paths of development. Thus, any one of the cells of two-, four-, or eight-celled embryos can be destroyed with a needle, and a normal mouse will still develop. Also, single cells from embryos with up to eight cells will develop in culture into 64-celled embryos of normal appearance. If such single cells are fused with 32-celled embryos, they participate in normal development. Again, if two eight-celled embryos are fused, they will develop within the uterus of a foster mother into one normal mouse.

The latter peculiarities of mouse embryos are shared with human embryos and allow, for better or worse, spectacular manipulations. For example, eggs can be removed from a woman and fertilized in a culture dish. If either parent's family has a history of a certain inherited disease, a single cell can be removed from each of the early embryos to discover whether it has acquired the defective gene that causes the disease. If not, when the embryo is implanted back into the mother, it should develop into a healthy child who cannot transmit the disease. The single cell could also be tested for male or female sex chromosomes and a male or female embryo, as desired, selected for implantation.

In eight-celled mouse embryos, regions of different composition can be detected within cells, and asymmetric division of these regions between daughter cells appears to be the origin of different cell fates.[15,16] These differences arise in cells that are exposed to the outside of the embryo: their outer surface becomes covered with finger-like projections, and the cytoplasm on this side of the cells has a distinctive appearance. When these cells divide, the plane of division usually separates the outer from the inner cell surface, and a 16-celled embryo is formed with an outer layer of cells of different composition and structure from those embedded in the center. Experiments suggest that molecules associated with the outer cell surface direct the outer cells toward forming the trophectoderm, and, in their absence, cells of the inner cell mass are formed. Thus, when outer cells of the 16-celled embryo divide, the plane of division occasionally does not separate the outer and inner cell surfaces but is at right angles to this, to give rise to two outer cells. As a result, more than half of the new cells contain molecules associated with the outer surface.

Experiments in which single cells were marked showed that somewhat more than half of the progeny of the outer cells become trophectoderm. This is consistent with the belief that these are cells that inherit the distinctive cytoplasm. The spherical inner cells contribute only to inner cell mass. However, if one of these cells is implanted on the outside of an embryo, it becomes asymmetric and contributes progeny to the trophectoderm. This transformation is blocked by drugs that inhibit gene activation, suggesting that it requires the formation of new molecules (although the details are unknown).

The transformation can be reproduced and studied in cells that are taken from eight-celled embryos before the transformation occurs and then made to adhere to one another. When two of these cells are associated, the distinctive cytoplasm and the projections on the cell membrane appear after 7 hours on the tip of each cell farthest from their point of contact. When three or four cells are associated in a chain, these changes occur only on the tips of the two outer cells. Adhesion of the cells largely depends on the cell adhesion molecule L-CAM (liver cell adhesion molecule). If this is inactivated, the change is delayed and the axes of asymmetry are misplaced, suggesting that this molecule is involved in initiating the transformation at the opposite pole of the cell. As described in Chapter 4, further differences in composition of the cells of mouse embryos are soon induced by signals released by one cell and received by another.

REFERENCES

1. Hamburger H. *The Heritage of Experimental Embryology: Hans Spemann and the Organiser.* New York: Oxford University Press, 1988.

2. Hörstadius S. The mechanics of sea urchin development studied by operative methods. *Biol Rev Cambr Philos Soc* 1939;14:132–179.
3. Hörstadius S. *Experimental Embryology of Echinoderms*. Oxford, UK: Oxford University Press, 1973.
4. Cameron RA, Davidson EH. Cell type specification during sea urchin development. *Trends Genet* 1991;7:212–218.
5. Strome S. Asymmetric movements of cytoplasmic components in *Caenorhabditis elegans* zygotes. *J Embryol Exp Morphol* 1986;97(Suppl):15–29.
6. Horvitz HR, Herskowitz I. Mechanisms of asymmetric cell division. *Cell* 1992;68:237–255.
7. Edgar LG, McGhee JD. Embryonic expression of a gut-specific esterase in *Caenorhabditis elegans. Dev Biol* 1986;114:109–118.
8. Wood WB, Edgar LG. Patterning in the *C. elegans* embryo. *Trends Genet* 1994;10:49–53.
9. Illmensee K, Mahowald AP. Transplantation of posterior pole plasm in *Drosophila*. Induction of germ cells at the anterior pole of the egg. *Proc Natl Acad Sci USA* 1974;71:1016–1020.
10. Sander K. Specification of the basic body pattern in insect embryogenesis. *Adv Insect Physiol* 1976;12:125–238.
11. Kageura H, Yamana K. Pattern regulation in defect embryos of *Xenopus laevis. Dev Biol* 1984;101:410–415.
12. Nieuwkoop PD. Inductive interactions in early amphibian development and their general nature. *J Embryol Exp Morphol* 1985;89(Suppl):333–347.
13. Gurdon JB, Mohun T.J, Fairman S, Brennan S. All components required for the eventual activation of muscle-specific actin genes are localized in the subequatorial region of an uncleaved amphibian egg. *Proc Natl Acad Sci USA* 1985;82:139–143.
14. Woodland HR. Identifying the three signals. *Curr Biol* 1993;3 27–29.
15. Johnson MH, Ziomek CA. Induction of polarity in mouse 8-cell blastomeres: specificity, geometry and stability. *J Cell Biol* 1981;91:303–308.
16. Johnson MH, Chisholm JC, Fleming TP, Houliston E. A role for cytoplasmic determinants in the development of the mouse early embryo? *J Embryol Exp Morphol* 1986;97(Suppl):97–121.

FURTHER READING

Davidson EH. How embryos work: a comparative view of diverse modes of cell fate specification. *Development* 1990;108:365–389.
Greenwald I, Rubin GM. Making a difference: the role of cell-cell interactions in establishing separate identities for equivalent cells. *Cell* 1992;68:271–281.
Gurdon JB. The generation of diversity and pattern in animal development. *Cell* 1992;68:185–199.

CHAPTER FOUR

SIGNAL MOLECULES INDUCE CHANGES IN CELL COMPOSITION THROUGHOUT VERTEBRATE DEVELOPMENT

Chapter 3 explained that differences in cell composition within an embryo arise in two ways: by asymmetric division of the contents of a parent cell and in response to a signal received by a cell from without. Most changes during development originate from inductive signals transmitted between closely aligned cells. However, late in the development of vertebrate embryos, some changes in cell composition are induced by hormones that are signal molecules carried in the blood. Thus, thyroid hormones in a tadpole induce changes in cell composition that initiate metamorphosis to a frog, including loss of the tail; and a rise in concentration of testosterone in a human embryo causes changes in cell composition that lead to the formation of male sex organs.

HANS SPEMANN REVEALED THE IMPORTANCE OF INDUCTIVE SIGNALS IN VERTEBRATES

Our understanding of inductive signals arose mainly from experiments done early in the 20th century. They were largely initiated by Hans Spemann from 1914 onward, first in a Kaiser Wilhelm Institute in Berlin and later in the University of Freiburg.[1] He worked largely with newts (*Triturus*) and salamanders, but it has since been shown that embryos of frogs, chicks, and mice give rise to essentially similar results. Remember that the fertilized egg of an amphibian first divides to form a hollow sphere of cells, the blastula, and that this invaginates to form the gastrula. Spemann studied the effects of transplanting groups of cells from one region of a gastrula to another, using surgical instruments that he had developed. The selected group of cells was sucked into a narrow glass tube and cut off around the orifice. Circles of equal size were taken in this way from different positions on two embryos, and each circle deposited in the hole left by the other and held in position for an hour until it had healed. The transplants were made visible by using two embryos with different shades of brown pigmentation.

Spemann first interchanged cells from the region of the embryo that later forms nerve cells with those from the region that later forms skin cells. With early gastrulae that had only recently invaginated, the transplanted cells developed like their new surrounding cells to give rise to a normal newt (**Fig. 4-1B**). Spemann concluded that the transplanted cells had been directed along this abnormal path of development by signals received in their new position. But with late gastrulae, the transplants developed into islands of nerve tissue in a sea of skin cells, or vice versa (Fig. 4-1B). Although the transplanted cells from early and late gastrulae appeared identical, their course of development clearly had been irreversibly **determined** during this time by signals that differed between upper and lower sides of the embryo.

When a blastula invaginates to give rise to a gastrula, it forms an indentation named the **blastopore**. This is said to have upper and lower **lips**, and mesoderm cells originate within its upper lip (see Fig. 1-10). Spemann had observed that the upper lip of a gastrula is formed from the region opposite the point of sperm entry, which in newts is conveniently marked by a **gray crescent** caused by dispersion of pigment granules. He had also found that when the cells of a two-celled embryo are separated, only the cell that contains the gray crescent develops into a normal adult. These and other observations suggested to Spemann that the upper lip is the source from which signals for the total development of the embryo originate, and he later named it the "organisator" (translated into English as "organizer"). This led him to do a famous experiment with a spectacular result. He excised the upper lip from an early gastrula and implanted it into the opposite side of another early gastrula. The transplanted upper lip did not develop like the surrounding epidermal cells and hence had already undergone an irreversible change. Instead, it caused an almost complete second embryo to form, so giving rise to Siamese twin tadpoles, presumably formed by a series of inductions initiated by the upper lip (**Fig. 4-2**).

The muscle cells of this second embryo clearly were formed from mesoderm cells of the transplanted upper lip. Spemann suggested two possibilities for the origin of the nerve cells. One was that they were formed from adjacent epidermal cells of the host embryo in response to a signal emitted by mesoderm cells of the transplanted upper lip. This was in agreement with a theory that the nerve cells of the spinal cord and brain are normally formed from epidermal cells of the ectoderm on the upper side of the embryo, these cells being formed in response to signals released by mesoderm cells that move forward beneath them during the formation of the gastrula. But Spemann preferred an alternative theory: that the signal is emitted not by

A EARLY GASTRULAE (SECTIONS)

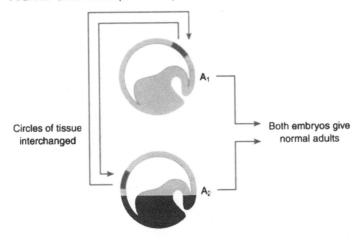

Circles of tissue
interchanged

A₁

A₂

Both embryos give
normal adults

B LATE GASTRULAE (SECTIONS)

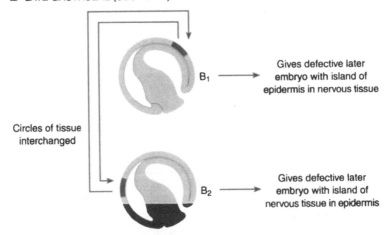

Circles of tissue
interchanged

B₁

B₂

Gives defective later
embryo with island of
epidermis in nervous tissue

Gives defective later
embryo with island of
nervous tissue in epidermis

FIGURE 4-1 Interchange of small circles of apparently identical epidermal tissue from two identical early gastrulae of newt (**A**) and two identical late gastrulae (**B**).

mesodermal but by epidermal cells of the dorsal lip, being transferred forward along the epidermis of the embryo from cell to cell and that this induces nerve cells. In Spemann's words, "the center for differentiation of the neural plate, at the beginning of gastrulation, is contained in the ectodermal part of the blastoporal lip and that differentiation spreads forward purely in ectoderm." If this was correct, the transplanted upper lip could not transmit this signal to the

EARLY GASTRULAE OF NEWT

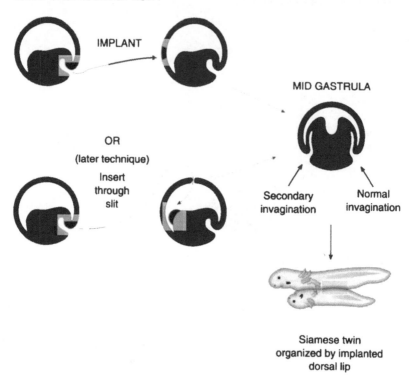

FIGURE 4-2 Transplantation of the dorsal lip of a blastopore from early gastrula into lower region of an identical gastrula.

host; in fact, the signal could be transferred only along epidermal cells of the transplanted lip, and these would be the source of any nerve cells that were formed.

Unfortunately, in this experiment differences in pigmentation did not show under the microscope, and it was unclear which cells came from the transplant and which came from the host. However, in 1924, the experiment was repeated by Hilde Mangold, a student of Spemann, on two species of newt (*Triturus cristatus* and *T. taeniatus*) whose pigmentation could be distinguished, and the nerve cells were found to be formed from the host embryo and not from the transplanted upper lip.[1] Also, a new and simpler technique in experiments confirmed that mesoderm cells of the transplant were the source of the inducing signal. The transplant was not grafted into the host embryo, but inserted through a slit into its central cavity where it became attached to the inner surface (see Fig. 4-2). When the upper lip of a blastopore was inserted in this way, an almost complete second embryo was again formed. The result was similar when mesoderm cells dissected from an upper lip were inserted, whereas the epidermal cells had no effect. Nerve cells were again shown to be formed from cells of the host.

That nerve cells are normally formed from epidermal cells in response to signals from underlying mesoderm was also supported by experiments of Johannes Holtfreter, a former student of Spemann's. He found that if the membrane that surrounds a blastula was removed before culture, the part of the blastula that normally invaginates at gastrulation now evaginated to form an **exogastrula** with a separate sac attached by a narrow stalk to the remainder of the embryo (**Fig. 4-3**). The extruded sac contained the mesoderm cells that normally move forward beneath the dorsal epidermal cells of the embryo. Numerous tissues except nerve tissue developed in this sac, including muscle, gut, and kidney. But the remainder of the embryo underwent little change and contained no nerve cells "even though the surface connection of the ectoderm with the upper blastoporal lip remained intact for a long period."[1] Holtfreter concluded, "the notion of an inducing stimulus progressing forward from the blastopore lip must be considered as invalidated."[1] Spemann never quite agreed. He still thought this might be an ancillary process prevented by the evagination, and recent work suggests that he was right. Although criticized by some for attributing the development of the whole embryo to the action of an ill-defined "organizer," Spemann's work largely led to our present understanding of inductive signaling during development, and he was awarded the Nobel Prize in 1935.[2]

NEURAL INDUCTIONS

The development of a vertebrate embryo clearly requires a series of signals to pass between cells that are temporarily and closely aligned during the continuous movement of cell layers. These interactions are known as **embryonic inductions**, and in the frog they first occur in the blastula to give rise to mesoderm cells (see Chapter 3). One induction that could be singled out

Ectoderm

Mesoderm (muscle and other tissues)

Endoderm

FIGURE 4-3 *Xenopus* exogastrula.

from the complex implications of Spemann's experiments was **neural induction**, the formation after gastrulation of nerve cells from epidermal cells on the upper part of the embryo in response to signals from the now underlying mesoderm cells. Similar inductions were soon demonstrated in embryos of several other species, including rabbits, chicks, and certain fish.

The nervous system has a complex structure, and different regions have cells of different composition. Spemann showed that formation of different regions requires different signals from mesoderm cells. He excised the upper lip of the blastopore from a newt embryo and separated the parts that migrate first and last through the blastopore during gastrulation. When these parts were grafted into gastrulae, they induced the formation of nerve cells that were characteristic, respectively, of the front and rear of the body. He also cut away newly induced nervous tissue from the outside of a newt gastrula to expose the underlying mesoderm and divided this into four equal portions from front to rear. Each portion was inserted into the cavity of a blastula, and each induced the formation of nerve cells, but the type of cell differed. The first three portions induced cells of the forebrain, midbrain, and hindbrain, respectively, whereas the fourth portion induced cells of spinal cord. These results were repeatedly confirmed by a new technique: each portion of the mesoderm was enclosed in a sandwich of epidermis from an early newt embryo, and the whole was incubated in a salt solution. The epidermal cells were again transformed into nerve cells, the type corresponding to the position along the embryo from which the mesoderm was taken. Without mesoderm, the epidermal cells merely multiplied to give an undifferentiated clump.

It appeared that signal molecules released by mesoderm might be easily identified by these simple techniques, and, in the 1930s, an outburst of experiments followed throughout the world. Spemann initiated these by showing that neural induction by the excised upper lip was little altered if it was first killed by heating or drying. He then found that many other tissues, including kidney from adult mice, would induce epidermal cells to become nerve cells, provided that the tissue was first killed. It was concluded that signal molecules normally released only by mesoderm were also liberated from these tissues by killing them. Many biologists then tested the ability of tissue extracts and pure compounds to induce the formation of nerve cells when inserted into the cavity of a blastula or sandwiched between isolated epidermis. To their consternation, many tissue extracts did this, as did many pure compounds including sterols and methylene blue dye. Moreover, newt epidermis would sometimes form nerve cells spontaneously—supposedly because the natural signal was being unmasked by maltreatment of the tissue. As a consequence, the search for the natural signals largely halted in confusion. The explanation of these findings is still unknown, although it is now clear that the signals for many embryonic inductions are minute quantities of protein growth factors that would have been impossible to identify with the techniques available at that time.

In spite of this confusion, Sulo Toivonen and Lauri Saxen[3] began a series of careful experiments in 1938 in Finland, which yielded consistent results, although their relevance is still unclear. They observed that when the hind end of the developing nervous system of a newt embryo was implanted into a gastrula, it induced nerve cells of the spine. However, if the tissue was first killed by heating it to 60° C, it induced brain. Acting on this clue, the researchers made a thorough study using a bizarre range of tissues that induced nerve cells. Liver of viper, guinea pig, and perch induced cells of the forebrain in over 80% of the gastrulae into which they were implanted. They also induced hindbrain cells in less than 20% and spine cells in none. In contrast, kidney of perch induced spine cells in 80% of the gastrulae, and forebrain cells in less than 5%. Kidney of guinea pig and jay were intermediate in their effect. Heating all these tissues before implantation increased the frequency with which cells of the forebrain were induced. When increasing amounts of unheated tissue were mixed with heated tissue, there was a progressive shift toward induction of the types of cells found toward the rear of the embryo. It appeared that two kinds of inducing molecules are active: a heat-sensitive molecule, which induces spinal cord cells when acting alone, and another molecule stable to moderate heating, which induces forebrain cells.

From these and further experiments, Saxen and Toivonen devised a theory of how the nervous system is first induced, which remains the basis of many theories of neural induction. The theory proposed that the first mesoderm cells to migrate through the blastopore lay down a uniform concentration of the heat stable inducer, which initiates the formation of forebrain cells in the overlying ectoderm. Its action is modified to give rise to the correct types of cell toward the rear by a rising concentration of the heat-sensitive inducer produced by mesoderm cells that migrate in later. This theory involved an idea, which rose to prominence in the 1930s, that a group of identical cells can be induced to form different types of cell by different concentrations of a single kind of molecule and that a range of cell types is often induced in embryos by a concentration gradient of such a molecule. (A molecule of this type is now called a **morphogen**.[4]) This mechanism remained unproved for over 50 years, but, as will be seen in Chapter 6, it has been shown to act in early *Drosophila* embryos. How widespread the mechanism is in embryonic development remains unclear. Recently, interest has been renewed in discovering the molecules that cause neural induction. (The progress that has been made is discussed in Chapter 7.)

INDUCTION OF EYES, FEATHERS, AND OTHER BODY STRUCTURES

The induction of mesoderm cells and nerve cells in the early vertebrate embryo is followed by a complex network of inductions that are largely responsible for producing cells of the correct composition in the correct position at each stage of development. Many of these inductions were studied in elegant experiments 50 or more years ago, but precise details of the signals involved and of the molecular reactions they induce in the recipient cells are only now beginning to be discovered. There has recently been an upsurge of interest in these aspects now that biochemical discoveries and new techniques promise success.[5,6]

Much work was done on the formation of the lens of the eye. The eye largely arises from cells of the brain that grow out to form the **optic cup**. The optic cup comes to lie against the epidermis of the head, and the lens is formed from this epidermis, which bulges inward and finally detaches as a hollow sphere that fits neatly into the cup. An experiment carried out by Ross Harrison at Yale University around 1920 appeared to show that the epidermis is induced to form the lens solely by a signal released from the optic cup. Harrison transplanted the developing optic cup of a frog embryo whose lens had not yet formed, to beneath the epidermis of its flank; and he also replaced the epidermis over the optic cup of a similar embryo with the epidermis from its flank. The epidermis over both cups formed the lens.

However, conflicting results soon arose. In some species of frog, the lens did not form when the epidermis overlying the cup was replaced by the epidermis from other regions, and the lens formed in the overlying epidermis even after removal of the optic cup. Further experiments showed that if it is to form lens, the epidermis from these frogs must first receive signals from other tissues. The requirement for these signals was found to differ among species, which explained the conflicting results. During the tissue movements of gastrulation, epidermis that finally lies over the optic cup and forms the lens passes over future endoderm and then over future heart mesoderm. The epidermis that would not form the lens could be made to do so if it is preincubated with excised fragments of these tissues before grafting over the optic cup. Clearly, these two tissues, as well as the optic cup, contribute to lens induction, and their relative importance differs among species. The signals they emit appear to be identical, since the effects of all three tissues are additive. Like the lens, the inner ear and inner tissue of the nose both form from the epidermis that overlies the brain. This epidermis again requires signals both from the underlying brain and from the endoderm and mesoderm that underlie it during the earlier tissue movements of gastrulation.

Other inductions during the development of the eye have been demonstrated. If a developing lens from a large species of newt is grafted against the optic cup of a smaller species (or vice versa), the optic cup grows faster (or slower) than usual. The lens appears to send signals back to the optic cup, and these signals regulate its growth. The **cornea**, the transparent outer window of the eye, is formed from the epidermis that covers the lens. If the lens is transplanted beneath the epidermis on another part of a frog embryo, this epidermis becomes transparent. It appears that a signal from the lens induces the overlying epidermis to become cornea. Also, if the cornea is removed, eyelids do not form, suggesting that a signal from the cornea is needed.

Feathers of birds and the scales on their legs are formed from epidermal cells in response to different inductive signals received from underlying mesoderm cells. Thus, if the mesoderm from a part of a chick embryo, which later forms feathers, is inserted beneath the epidermis, which normally does not, this epidermis later forms feathers. If the mesoderm from the upper leg is inserted beneath the epidermis on the wing, this epidermis later forms leg feathers. If the mesoderm from the foot is cultured with epidermis from the back, this epidermis forms scales. Conversely, mesoderm from the back with epidermis from the foot forms feathers. Similar experiments show that the fins and gills of amphibians, the beaks of birds, and the hair, teeth, and salivary glands of mammals arise from ectoderm cells that receive signals from underlying mesoderm.

The development of the pancreas, lung, and liver has been studied in mice and other animals, and all originate from endoderm cells that receive inductive signals from adjacent mesoderm. Signals from the mesoderm also largely induce the formation of cells that arise from the mesoderm itself. For example, the kidney is composed of several types of cells—all derived from mesoderm cells. Many experiments in which parts of the developing kidney of birds and mammals are removed have shown that essential signals are repeatedly sent from one kind of cell to another.

HARRISON'S EXPERIMENTS ON LIMB FORMATION

The development of the legs of amphibians and the wings and legs of chicks has been studied in great detail, and such experiments are described in the next two sections. Experiments were initiated around 1918 at Yale University by Ross Harrison, who appears to have been the only contemporary embryologist whom Spemann considered of similar stature to himself. Spemann had hoped that he and Harrison would share the Nobel Prize, but Harrison never received the award. Harrison worked with newts, in which the first visible sign of leg formation is the multiplication of four groups of mesoderm cells—two on each side of the embryo—soon after gastrulation. The cells push out the overlying ectoderm to form **limb buds**, which eventually form the forelegs and hind legs. These mesoderm cells appear to be differentiated from their neighbors by inductions during gastrulation. Harrison showed that legs can be induced at any point along the side of the embryo by implanting fragments of the developing ear or nose beneath the mesoderm cells, although these distant organs cannot be the normal source of the required signals. These implants induce forelegs along the front half of the embryo and induce hind legs along the rear half, showing that differences have already been induced between front and rear mesoderm cells.[7]

Harrison next made another interesting discovery. One might expect that development of an organ would result merely from a series of inductions in each of which a group of identical cells is divided into correctly placed subgroups of different composition. But Harrison showed that the process is more complex, since some groups of induced cells can compensate for change in size of the group and appear to act in a coordinated way. Such groups were called **fields**.[8] Harrison excised the mesoderm from limb buds of early embryos by inserting scissors through the ectoderm and withdrawing the excised tissue through the hole. When he removed all the cells that normally give rise to the leg, the surrounding mesoderm cells multiplied to heal the wound, and the embryo later formed a normal leg. Hence, the neighboring cells that do not normally form leg now did so. However, the cells with this ability were limited in extent, for when Harrison excised the same tissue together with an equal amount of surrounding tissue, the wound healed but no leg formed. When he removed half of the cells destined to form the foreleg and inserted them under the ectoderm elsewhere on the embryo, normal forelegs developed in both places. When Harrison excised the mesoderm cells from one limb bud and inserted them beside those of another bud, the two groups often fused to form one normal leg.

Other researchers showed that if a limb bud is slit vertically and the halves are prevented from rejoining by insertion of a membrane, then each half may form a complete leg. Induction of similar coordinated groups of cells is an early step in the development of several other organs in amphibiana, including gills, eyes, ears, nose, and heart. The size of each group of cells is small, less than 1 mm (50 cell widths) in diameter. The cells in a field appear to be in continuous communication with each other: if some of the cells are excised or more cells are added, the cells throughout change their fate so that an organ of normal structure is still formed. How this occurs is unknown. It has been suggested that there is a gradient in concentration of a signal molecule (a morphogen) across the field and that the behavior of each cell is determined by the local concentration, which will change when the field is disturbed (see also Chapter 4). An alternative suggestion is that when the field has its correct structure, all adjacent cells are compatible with one another, and when the field is disturbed, incompatible cells become adjacent. This stimulates division until compatibility is restored.

Harrison also made other important discoveries. Before describing them, I must clarify the asymmetric structure of vertebrate limbs. This is best illustrated by the human arm held out sideways with the palm down. Three lines (axes) at right angles to one another can be drawn through the arm, with the structure differing at either end of each axis. One runs from shoulder to fingertips (proximal-distal axis); another from the thumb in front to the little finger at the rear (anterior-posterior axis); and the third from above to below (dorsal-ventral axis). Because the structure differs at either end of each axis, no plane of symmetry can be drawn through a limb. Also, the right limb is the mirror image of the left.

Harrison excised early limb buds from newt embryos and grafted them back after rotation. If the hind limb bud was excised before the end of gastrulation, rotated 180 degrees, and replaced, a normal leg developed. But if the operation was performed a little later, the leg had the anterior-posterior axis reversed (equivalent to a human arm, when held out sideways with the palm down, having the little finger at the front and the thumb at the rear). If performed a little later still, the leg had both anterior-posterior and dorsal-ventral axes reversed (equivalent to a human arm rotated 180 degrees so that the elbow faces forward). These experiments implied that development along the three axes is directed by three separate sets of inductions.

THE ORIGIN OF CELLS THAT LAY DOWN THE STRUCTURE OF A CHICK'S WING

More recent studies of limb development have been made mostly on chick embryos because they are more convenient for experiments. Their limb buds are large, and since the embryo lies on the yolk of the egg with its right wing bud uppermost, it can be operated on with little disturbance. A window is cut in the eggshell above the bud, and after the operation, the shell is replaced and sealed with wax.

The initial development of chick wings and legs appears to be similar to that of newts, and most later work has been directed toward discovering the source of signals along the three different axes that give rise to the asymmetric structure and arrangement of bones as the legs and wings grow outward. The epidermal cells covering the tips of growing chick limb buds are tall and thin and form a narrow **apical ectodermal ridge**, running from front to back. Beneath epidermal cells of the ridge is a thin layer of rapidly dividing mesoderm cells, the **progress zone**, which largely provides the new cells needed for growth. It was found that the apical ridge can be removed from a wing bud by inserting a curved glass needle beneath it and tearing it away. When this was done, neighboring epidermis healed over the wound, but cell division in the progress zone largely ceased and the wings that developed were defective (**Fig. 4-4**). The wings had perfect structures adjacent to the body but lacked those toward the end of the wing to an extent that depended on when the apical ridge had been removed.

A WING BUD

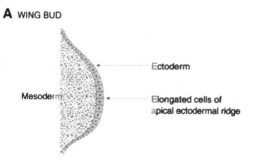

Ectoderm

Mesoderm

Elongated cells of apical ectodermal ridge

B NORMAL SKELETON OF CHICK S WING

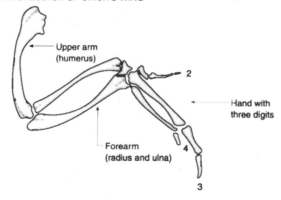

Upper arm (humerus)

2

Hand with three digits

Forearm (radius and ulna)

4

3

C TRUNCATED SKELETON FORMED WHEN APICAL RIDGE IS REMOVED IN MID DEVELOPMENT

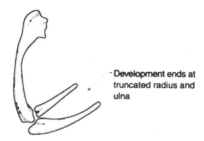

Development ends at truncated radius and ulna

FIGURE 4-4 Example of truncated skeleton of chick's wing formed when apical ectodermal ridge is removed during development.

Structures along the length of the wing are colloquially named after homologous parts in a human arm, namely, pectoral girdle, shoulder, upper arm, forearm, wrist, and hand. If the ridge was removed very early, only part of the shoulder near the body formed correctly. If removed a little later, the wing ended somewhere along the upper arm; if later still, somewhere along the forearm; and if later still, somewhere along the wrist or hand. It was concluded that cells leaving the progress zone over several days in turn provide successive structures along the length of the wing. Apart from maintaining cell division in the progress zone, the apical ridge did not appear to determine the wing structure, because if the ridges were interchanged on two wing buds of different ages, their development was unaffected.[9-11]

The latter conclusions were confirmed by further experiments in which the progress zones were interchanged between pairs of chick embryos of different ages. The right wing buds of two embryos, one 3 and one 5 days old, were exposed by removing a small piece of eggshell. A slice about 0.4 mm thick was cut from the tip of each wing bud and consisted of the ectodermal ridge plus the underlying progress zone. Each excised tip was then attached to the stump of the other wing bud by platinum pins, and the eggshell was replaced. The wings were removed when each embryo was 10 days old and were then examined under the microscope. The wing formed from the 5-day stump plus 3-day tip had certain structures repeated more than once along its length (**Fig. 4-5**). It consisted of shoulder, upper arm, forearm, upper arm, forearm,

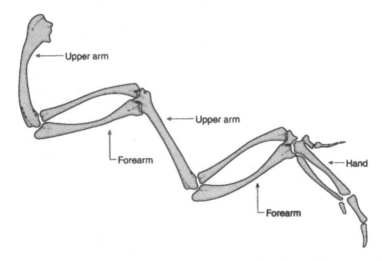

FIGURE 4-5 Result of exchanging wing bud tips between chick embryos of different ages. Here, the tip from a 3-day embryo was grafted onto the stump from a 5-day embryo, and the wing developed as shown.

and hand. The wing formed from the 3-day stump plus 5-day tip had structures missing. This wing consisted of shoulder, upper arm, and hand; the forearm was absent. These experiments confirmed that cells continually leaving the progress zone during growth of the wing in turn form the successive structures along its length. They also proved that the cells are not merely laid down in successive strata and then induced to form the wing. In contrast, when the mesoderm cells leave the progress zone, which is only about 20 cell diameters deep, they must already have acquired a composition that directs them toward forming a particular part of the wing. These mesoderm cells appear to direct wing structure by forming a scaffold of bone on which muscle cells, which migrate in from outside the wing, and nerve cells position themselves secondarily.

THE SOURCE OF SIGNALS THAT INDUCE THE STRUCTURE OF A CHICK'S WING

Harrison's experiments suggested that the structure along each of the three axes of a limb is determined by a separate set of inductions. Experiments described in the last section show that, in a chick embryo, when cells leave the progress zone, they have already received all signals that enable them to lay down the correct three-dimensional structure of the wing. However, the origin of only one of the three sets of signals that are required has been discovered, namely, those that act along the anterior-posterior axis (from thumb to little finger).[9-11]

The first important clue to the origin of these induction signals came from an experiment designed for another purpose. Mesoderm cells from the rear of the right wing bud of a chick embryo were inserted beneath the tip of a similar bud. The purpose was to test the effect of these rear mesoderm cells, which are overlaid by thin ectoderm cells, on the thick ectoderm cells at the tip of the apical ridge. These thick cells became thin, but a more surprising and interesting change also occurred in this and subsequent experiments (Fig. 4-6A). The ectoderm cells in front of the transplanted mesoderm cells thickened to become like those of an apical ridge, and in front of the normal wing a second set of wing parts began to grow out from this region. A wing finally formed with the normal structure of the original wing close to the body but duplicated (usually with some abnormalities) toward the end and with a complete (or almost complete) extra set of digits ("fingers"). The younger the embryo at the time of transplantation, the greater the length of the wing that was duplicated. Only mesoderm cells from the rear of a wing or leg bud caused this duplication.

In this experiment, the arrangement of digits on the duplicated parts was normal. But when similar mesoderm cells were inserted at the front rather than

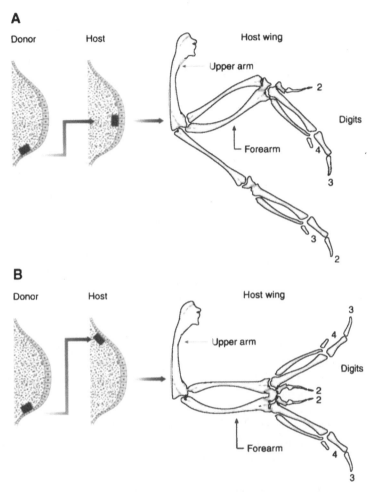

FIGURE 4-6 Examples of the effects on wing structure of transplantations of mesoderm from the rear of a donor wing bud into the wing bud of a host chick.

at the tip of a right-hand wing bud, the result was different (Fig. 4-6*B*). A second set of wing parts grew out from behind the inserted cells rather than in front of them, although they were still in front of the normal wing. Although not all digits were present on the additional wing, those that were present were abnormally arranged. The chick's wing, from front to back, has three digits denoted No. 2 ("thumb"), No. 3, and No. 4 ("little finger"). (Nos. 1 and 5 of the typical vertebrate limb are absent.) In the additional wing, the normal order of digits from front to rear was reversed. For example, in one of the duplicated wings the order from front to back was No. 4 and No. 3, followed by Nos. 2, 3, and 4 of the normal wing, all placed the correct side up. In later experiments, all three digits were duplicated, as shown in Figure 4-6B. It was as though the normal right-hand wing had the end of a left-hand wing attached to its front. (The reader can easily make the structure clear by laying his or her fingers flat on a table with the two forefingers side by side.) Hence, in the additional wing, the structure along the anterior-posterior axis was reversed, whereas that along the other two axes was normal.

In another further series of experiments, mesoderm from the rear of one wing bud was again inserted beneath the tip, or beneath the front, of another. But after the recipient bud had healed, its own rear mesoderm was removed and discarded. This time, only one wing was produced: a normal right-hand wing in front of transplants at the tip and, behind transplants at the front, a left-hand wing with its normal front pointing toward the rear (i.e., with its structure along the anterior-posterior axis reversed). Many similar experiments were done on leg buds of chick embryos with similar results. It was concluded that the mesoderm at the rear of a limb bud (the **zone of polarizing activity**) is the source of signals that determine structure along the anterior-posterior axis within cells of the adjacent progress zone, those cells that are the first to receive the signal becoming the posterior of the wing.

As for structure along the length of the limb (proximal-distal axis), it was suggested some years ago that this structure is determined by the increasing time that successive cells spend in the progress zone and arises not from inducing signals but from changes in molecular composition that occur with each cell division. This idea was made improbable by an experiment in which cells that had emerged from the progress zone of a chick embryo early in limb growth had their fate changed to that of cells that emerge late by returning them to the progress zone at this stage. Mesoderm cells from the base of a leg bud, which normally form thigh, were implanted beneath the tip of a mature wing bud in which structures at the end of the wing were being induced. A normal wing developed except that at its tip were toes and claws typical of a foot. Hence, leg cells that had a molecular composition directing them to form the thigh had that composition changed to one directing them to form toes. This strongly suggests that fate is not determined by the number of cell divisions that occur beneath the ridge, but by signals from outside the cells that change with time. When a layer of wing mesoderm cells intervened between the implanted thigh tissue and the tip of the wing bud, the change did not occur: thigh feathers formed at the end of the wing instead of toes. This implied that the signals originate in the mesoderm cells at the tip of the limb bud.

Signals that induce dorsal-ventral development appear to be released by ectoderm, since rotation of the ectodermal covering of a limb bud 180 degrees could invert the structures along this axis. Recently, many more discoveries have been made about the signals that induce limb growth, as described in Chapter 7.

SIGNALING CELL POSITION

I have already mentioned that embryos appear to contain small coordinated groups of interacting cells, or fields. Simple experiments have also revealed the existence of much larger coordinated groups of cells in embryos and adults. They can be demonstrated most easily in insects in which excision of part of the body often stimulates vigorous regeneration, or regrowth, of the removed part. The first clue came about 30 years ago from experiments on a bug called *Rhodinus*.[12] As usual, the larva of this insect is segmented, and each segment is covered with a hard cuticle that is secreted by underlying epidermal cells. The larva grows with a series of moltings in which the old cuticle is shed and replaced by a newly formed one. The cuticle on each segment of the adult has parallel transverse folds or ripples. Square pieces of the cuticle plus underlying epidermis can be excised from one larva and inserted in a hole of similar size in the cuticle plus epidermis of another larva. The transplant heals, but after the next molt, the pattern of ripples on the new cuticle formed by the transplanted epidermis is sometimes changed.

If the square of tissue is transplanted without rotation into the same position on any segment of another larva, the pattern is unchanged. It is also unchanged if the new position is to the left or right of the old, but at the same distance from the front of the segment. However, if the transplant is rotated 180 degrees before insertion or inserted farther forward or back within the segment, then the pattern changes (**Fig. 4-7**). Clearly, when epidermal cells from different positions along the anterior-posterior axis of any segment are placed next to one another, an interaction occurs, which results in disruption of the epidermis and hence of the pattern on the cuticle that it secretes. In other insects, it has been shown that the disruption is caused by stimulation of cell division at the junction. Although all the epidermal cells look identical, there must be some molecular difference between them that causes division to be stimulated when cells not normally in contact along the anterior-posterior axis are placed together. The fact that insertion of a transplant sideways from its previous position does not affect the cuticle pattern does not necessarily mean that cell division is not also stimulated by a move in this direction. Cell division may occur without affecting the cuticle.

Experiments on the legs of adult cockroaches have confirmed the latter findings and proved that, in the cockroach, cell division is stimulated by movement of the epidermis in either of two directions at right angles. If

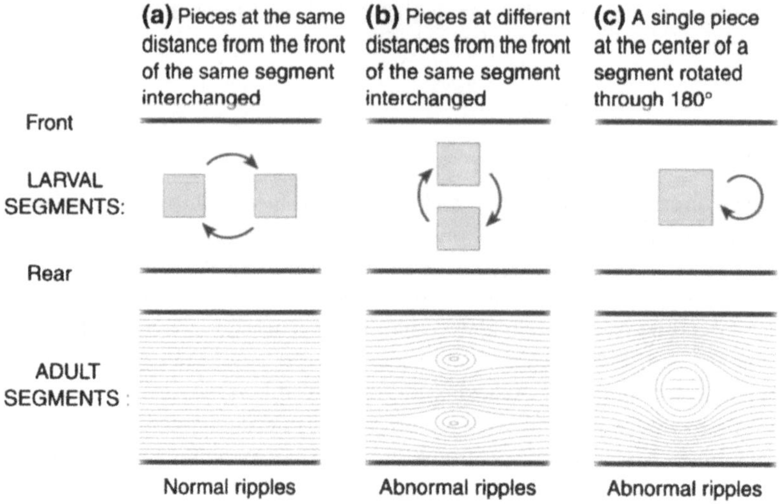

| | (a) Pieces at the same distance from the front of the same segment interchanged | (b) Pieces at different distances from the front of the same segment interchanged | (c) A single piece at the center of a segment rotated through 180° |

FIGURE 4-7 Effect of transplantation of pieces of cuticle within the larva on the cuticle of the adult *Rhodinus.*

longitudinal strips of cuticle plus the underlying epidermis are removed from identical positions on the upper legs of two cockroaches and interchanged without rotation, each heals to give an unblemished leg. The same occurs if the interchange is between similar positions on two different segments of the leg. However, if the strips are taken from the same position down the length of the leg but different positions sideways around its circumference, healing is abnormal in that cell division is activated along both longitudinal sides. Similar changes also occur when the epidermis from different positions down the length of the leg comes in contact (**Fig. 4-8**). If the upper leg of each of two cockroaches is cut across sideways at the same level and the end of each leg is grafted onto the stump of the other, each pair heals without growth to give rise to a normal leg. Healing without growth also occurs if the cuts are made across different segments of the leg at similar levels within each segment. But if the cuts are at different levels, cell division and growth occur at each junction, with both graft and stump forming new cells. When the junction is between the shorter end of the leg and the shorter stump, the leg grows to regain its normal length. When it is between the longer end and the longer stump, cell division gives a leg that is longer than normal. Here, the new growth is unusual in that the bristles on the cuticle point upward rather than downward.

A theory to explain these facts can be devised as follows:[13-15] cells in each segment of the leg are arranged in different subgroups of one or a few cells, the arrangement being identical in each segment. Adjacent subgroups differ in

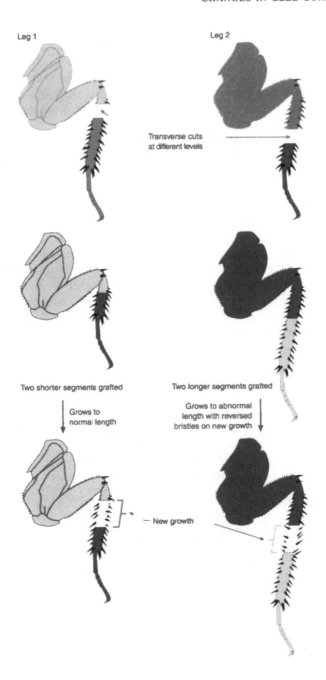

Leg 1

Leg 2

Transverse cuts
at different levels

Two shorter segments grafted

Two longer segments grafted

Grows to
normal length

Grows to abnormal
length with reversed
bristles on new growth

New growth

FIGURE 4-8 Effect of interchanging pieces of different length cut from the legs of two cockroaches.

having variations in two kinds of molecule on the cell surface. In each successive subgroup down the length of the leg, the first of these molecules is different and the second is the same. In each successive subgroup around its circumference, the first is the same and the second is different. If the variations in the first kind of molecule are lettered A, B, C, D, E, ... X, Y, Z and those in the second kind of molecule a, b, c, d, e, ... x, y, z, their arrangement in the mature leg can be illustrated (**Fig. 4-9**).

In the mature leg, all subgroups are compatible and cell division is not stimulated. (That is, in the illustration, each letter of the alphabet is adjacent to its correct neighbors.) When subgroups with one or both incompatible molecules are aligned by grafting, (i.e., some letters of the alphabet are absent between them), signals pass between them that induce cell division with the formation of all intermediate subgroups, a process called **intercalation**. The newly formed cells do not merely have the correct molecules on the surface, but they also have a composition and structure appropriate for their position. (The same result could conceivably be directed by two morphogen gradients at right angles to one another. However, it is difficult to believe that such gradients could survive the grafting operations, especially since severed fragments may be kept in culture for several days before regrafting.)

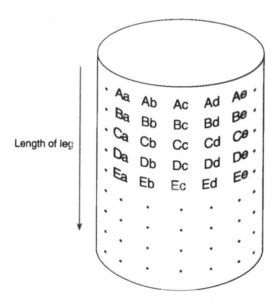

FIGURE 4-9 Hypothetical arrangement of cell subgroups on each segment of the leg of a cockroach.

Such an explanation is supported by the reversed orientation of bristles on the new growth between the long end of one severed leg grafted onto the long stump of another, as just described. Suppose that cells of the stump with molecular assembly H on their surface confront cells of the severed leg with assembly D giving rise to the sequence

$$\dots E, F, G, H, D, E, F, G, \dots$$

At the junction, H and D cells are incompatible and division is induced to give a sequence of compatible cells, namely,

$$\dots E, F, G, H, \underline{G, F, E}, D, E, F, G \dots,$$

when cell division ceases. However, cell subgroups along the new growth ($\dots \underline{G, F, E} \dots$) are in reverse order to those along the remainder of the leg. Hence, the orientation of the bristles is reversed.

Although differences between cells that relate solely to their position are most easily demonstrated in insects, they also appear to exist in amphibians whose severed legs can regenerate. If a leg of each of two adult newts is cut across, each at the same level, and the end of one grafted onto the stump of the other, the two heal without elongation. But if cut across at different levels and the shorter end is grafted onto the shorter stump, then cell division occurs at the junction until the leg attains its normal length and structure.

The significance of these facts is unclear, but it could be immense. These facts may be revealing the process that regulates normal growth to maturity: in immature tissues, adjacent cell surfaces may be incompatible, so that cell division is stimulated until all are compatible. It is in fact possible to interpret the facts of limb growth, described earlier in this chapter, as determined by interactions between adjacent cell surfaces rather than by diffusible signal molecules. The different arrangement of similar cells in different organs, such as the arm and leg, might result from the initial induction of different surfaces in these cells so that they undergo different numbers of cell divisions to maturity. No molecular differences between cells that relate solely to their position have yet been discovered, but an understanding of the molecular basis could conceivably enable the regeneration of human tissues with benefits to surgery. The topic is wide open and awaits further research.

REFERENCES

1. Hamburger H. *The Heritage of Experimental Embryology: Hans Spemann and the Organiser*, New York: Oxford University Press, 1988. (A fascinating personal account of the early days of experimental embryology in Spemann's laboratory.)

2. Gilbert SF, Saxen L. Spemann's organiser: models and molecules. *Mech Dev* 1993;41:73–89.
3. Saxen L, Toivonen S. The two gradient hypothesis in primary induction. The combined effect of two types of inductors mixed in different ratios. *J Embryol Exp Morphol* 1961;9:514–533.
4. Neumann C, Cohen S. Morphogens and pattern formation. *BioEssays* 1997;19:721.
5. Jacobson AG. Inductive processes in embryonic development. *Science* 1966;152:25–34.
6. Wessels NK. *Tissue Interactions and Development.* Menlo Park, CA: Benjamin/ Cummings, 1977.
7. Harrison RG. Experiments on the development of the fore-limb of *Amblystoma*, a self-differentiating equipotential system. *J Exp Zool* 1918;25:413–461.
8. Ingham PW, Martinez Arias A. Boundaries and fields in early embryos. *Cell* 1992;68:221–235.
9. Saunders JW Jr, Gasseling MT. Ectodermal-mesenchymal interactions in the origin of limb symmetry. In Fleishmajer R, ed. *Epithelial-Mesenchymal Interactions in Development,* 5th ed. Baltimore: Williams & Wilkins, 1968:78–98.
10. MacCabe JA, Saunders JW Jr, Pickett M. The control of the anteroposterior and dorsoventral axes in embryonic chick limbs constructed of dissociated and reaggregated limb-bud mesoderm. *Dev Biol* 1973;31:323–335.
11. Wolpert L, Lewis J, Summerbell D. Morphogenesis of the vertebrate limb. *Ciba Society Symposia* 1975;29:95–118.
12. Martinez Arias A. A cellular basis for pattern formation in the insect epidermis. *Trends Genet* 1989;5:262–267.
13. Bryant PJ, Bryant SV, French V. Biological regeneration and pattern formation. *Sci Am* 1977;237(No. 1):66–81.
14. Bryant SV, FrenchV, Bryant PJ. Distal regeneration and symmetry. *Science* 1981;212:993–1002.
15. Bryant SV, Muneoka K. Views of limb development and regeneration. *Trends Genet* 1986;2:153–159.

FURTHER READING

Huxley J, de Beer G. *The Elements of Experimental Embryology.* Cambridge, UK: Cambridge University Press, 1934.
Waddington CH. *Principles of Embryology.* London: George Allen and Unwin, Ltd, 1956.

CHAPTER FIVE

DIFFERENCES IN EMBRYONIC CELL COMPOSITION RESULT FROM ACTIVATION OF DIFFERENT GENES

The facts about animal development described so far in this book have largely been discovered by simple manipulations with tweezers, hairs, and razor blades, which produce visible results that are easy to understand. We now move down in scale by several dimensions to studies of the invisible—but no less real—molecular interactions that are the driving force of embryonic development. This, however, need not deter readers who have little knowledge of molecular biology. The chemical reactions of living cells are not difficult to understand and can be accurately represented by changes in the structure of molecular models used by research workers, such as those in which plastic spheres of various sizes represent atoms and rubber plugs represent chemical bonds. Our descent into the molecular realm begins with Mendel's very straightforward experiments on the development of pea plants, which led directly to our understanding that differences in the molecular composition of embryonic cells arise from the activation of genes with differences in molecular structure.

MENDEL FIRST DEMONSTRATED THE IMPORTANCE OF GENES IN DEVELOPMENT

The development of a fertilized egg into an adult organism is founded on the ability of cells to beget like cells, which is sometimes modified by the controlled introduction of differences in molecular composition. This cellular heredity is illustrated most simply by bacteria. For example, an *Escherichia coli* bacterium divides to give rise to two bacteria like itself. This mechanism might appear simple: at cell division, the component molecules and the structures formed from them are evenly distributed between the daughter cells. However, this explanation is inadequate, since the daughter cells are half the parental size and each kind of molecule must double in number before the next cell division. This replication of cell molecules is the basic problem of heredity. Thus, flower color in peas, which Mendel studied, results from cells of the flower forming pigment molecules. These molecules were not present in the pollen or egg cell that united to give rise to the fertilized egg. However, they must have been formed by molecules that were present in one or both of these germ cells and were exactly copied before each of the cell divisions that led to the flower. Our knowledge that these are self-copying DNA molecules stems from a sequence of experiments that derives from Mendel.[1]

Gregor Mendel[2] was born in 1822 in the Austrian Empire, the son of a peasant farmer. He was a distinguished student in high school, and, in 1843, he was admitted to the Altbrunn Monastery in Brunn as a member of the community of about 13 priests chosen for their intellectual ability. After his ordination in 1851, he was sent to Vienna University for 2 years to study mathematics, physics, and biology. He was taught by some of the leading scientists of the day, whose ideas influenced his subsequent experiments. From the time of his return to the monastery until he was elected abbot in 1868, he taught physics and biology in a large and distinguished secular high school in Brunn. In his spare time, Mendel studied inheritance in pea plants, of which he grew over 30,000 in the monastery garden with the help of the gardener.

Before Mendel announced the results of his experiments, there had been no satisfactory theory of how characteristics are transmitted from one generation of individuals or cells to another, in spite of numerous studies of **hybridization**, the crossing of plants or animals with differing characteristics. A popular theory was that of **blending inheritance**: both parents contribute to their offspring materials that blend together, the relative proportions determining its characteristics. But this theory was not generally accepted, since characters that appeared blended or lost in a first cross, such as black fleece in sheep, often reappear unchanged in subsequent generations. This suggested that inheritance is determined by the interaction of discrete particles. However, as Mendel noted, "... among all the numerous experiments made, not one has been carried out in such a way as to make it possible to determine the number of different forms under which the offspring of hybrids appear, or to arrange these forms with certainty according to their separate generations, or to definitely ascertain their statistical relations."[3] This, he determined to remedy.

Chemists of the 18th century discovered that chemical changes are founded on the interaction of particles (named molecules) because they made precise measurements of the quantities of compounds that react together rather than merely describing the appearance of the products. Similarly, Mendel proved that the transmission of characters is founded on the transmission of particles (now named genes) because he made precise counts of the plants that showed different characteristics.

Mendel chose pea plants for his experiments for two reasons: varieties could be bought from seed merchants with unambiguous differences in certain points of character, and also their flowers are usually self-fertilized and receive no pollen from other plants unless it was introduced experimentally. He chose strains of plants that differed in one or more of seven characteristics and maintained these differences through many generations of self-fertilization. The differing characteristics were round or wrinkled seed, yellow or green seed, red or white flowers, smooth or wrinkled pods, green or yellow pods, flowers distributed along the stem or bunched at the end, and stems of about 2 meters or about ½ meter. When he crossed plants that differed in one or more of these characteristics, one of the alternative forms always appeared in the hybrid to the exclusion of the other. For example, when pollen from round seeded plants was used to fertilize the flowers of wrinkled seeded plants or vice versa, the seeds that formed from these flowers were always round. Or, when red-flowered plants were crossed with white-flowered plants and the seeds that formed were grown into new plants, the plants all had red flowers. Mendel named round seed and red flowers **dominant** characters and their alternatives **recessive**. The other dominant characters were yellow seed, smooth pods, green pods, flowers along the stems, and stems of about 2 meters.

Mendel then grew large numbers of each of the seven kinds of hybrid seed into plants that he allowed to self-fertilize, and he later collected the seeds that formed. Some of these seeds, or the plants grown from them, showed the recessive forms of the characters that had disappeared in their parent. He counted the numbers of dominant and recessive forms and found a startlingly clear relation between them—a sure sign of a simple underlying mechanism. For example, he grew 253 hybrid round-seeded plants and allowed them to self-fertilize. Usually, both round and wrinkled seeds were found in the pods that formed on the plants. He collected 7324 seeds; 5474 were round and 1850 were wrinkled—a ratio of 2.96 to 1. From each of the six other kinds of hybrid plants, he obtained after self-fertilization the dominant and recessive forms in the ratios of 3.01, 3.15, 2.95, 2.82, 3.14, and 2.84 to 1. All these ratios are close to 3 to 1, and they usually approached it most closely when large numbers were counted.

Mendel also performed experiments to test whether different characters were inherited independently. For example, he had two strains of plants that had always bred true and differed in two characters: one gave rise to seeds that were both round and yellow, and the other gave rise to seeds that were both wrinkled and green. When these plants were crossed, the seeds of the hybrid plants were all round and yellow, as expected. These plants were allowed to self-fertilize. Would their round seeds always be yellow and their wrinkled seeds always be green? The answer was no. Out of 556 seeds, 315 were round and yellow, 101 wrinkled and yellow, 108 round and green, and 32 wrinkled and green. The ratio between these is 9.8 to 3.2 to 3.4 to 1. Mendel concluded that the characters are inherited independently and that the deviation from the expected ratio of 9 to 3 to 3 to 1 was due to chance.

Mendel presented these and other results, together with a theory to explain them, at a meeting of the Brunn Natural History Society, and his paper was published in 1866 in the Transactions of the Society (soon after the publication of Darwin's *The Origin of Species* in 1859). His theory was in effect made up of two parts. The *first* part of Mendel's theory suggested that the inherited characteristics that any pea plant reveals during its development can be subdivided into a large but limited number of unit characters. Plants that differ in their inheritance possess one or more of these unit characters in sharply different forms. One plant, for example, can have red flowers and another can have white. This revolutionary theory suggested how a precise analysis could be made of inherited differences that had previously been considered only in a descriptive way. It implies that the number of genetically different individuals in a species, though large, is limited. For example, genetically different humans result from the alternative forms of thousands of unit characters being combined in different ways. If all these ways were exhausted, any further individuals would be identical twins of those who had come before. In fact, such a particulate theory is necessary if inheritance and development are to be explained in molecular terms.

The *second* part of Mendel's theory suggested how the particular character of each plant is determined by its parents. He proposed in effect (not in so many words) that every egg and pollen cell of the pea plant carries for each unit character one **determinant** (later to be named a gene). The fertilized egg therefore carries two determinants for each character. During development, each pair of determinants is passed into all cells of the plant (or at least into the germ line cells and the cells in which they are required to act), and the form in which each character appears is decided by interaction of the members of each pair. For example, every egg and pollen cell carries a determinant for flower color. In strains of red or white plants that breed true, both egg and pollen will contain the same determinant—either for red or for white. When these plants are crossed, each fertilized egg, and later each cell of the flower, will contain one determinant for red and another for white. Moreover, since red is dominant over white, the plants will have red flowers. But half of the germ cells formed by these hybrid plants will contain a red determinant, and the other half will contain a white determinant. Hence, when these plants self-fertilize, "it remains," in Mendel's words, "purely a matter of chance which of the two sorts of pollen will become united with each separate egg cell."[3] Like the simultaneous tossing of two coins: there is one chance that two heads will fall together, to two chances that a head will fall with a tail, to one chance that two tails will fall together. Hence, after self-fertilization, approximately one-fourth of the fertilized eggs will contain two determinants for red, another fourth will contain two determinants for white, and the remaining half will contain one of each determinant. But only those with two determinants for white will give white-flowered plants, thus giving a ratio of red to white of 3 to 1. The red-flowered plants should be of two kinds. One-third on self-fertilization should give rise to only red-flowered plants, whereas the remainder should give rise to red and white in a ratio of 3 to 1. Mendel presented further results showing that this was so.

MENDEL'S RESULTS LED TO THE DISCOVERY THAT GENES ARE MADE OF DNA

Mendel's paper was widely circulated and read by a number of leading biologists, but none recognized its fundamental importance. He died in 1884 without the recognition he deserved. Mendel's loss was Darwin's gain. There would have been little support for Darwin's theory that one species can be transformed into another by the Natural Selection of small differences between individuals if it had appeared that these differences result solely from the reassortment of a limited number of fixed characters. Integration of the two theories awaited the discovery, early in the 20th century, that Mendel's unit characters can undergo almost endless change by the mutation of their determinants or genes and that each species carries a large reservoir of mutant genes on which selection can act.

In 1900, three botanists published findings similar to those of Mendel on peas and other plants. During their work, they had read Mendel's paper and acknowledged his priority. By this time, important advances had occurred in biological knowledge and thought, which made clear the importance of his work. During the 1880s, details of cell division had been clarified. The nuclear material was seen to condense into separate threads called **chromosomes**, which appeared to divide longitudinally to give rise to identical chromosomes that separated into daughter cells. August Weismann (1834–1914) had suggested around 1890 that chromosomes carry the hereditary material: "a substance with a definite chemical and, above all, molecular constitution."[1] He proposed that egg and sperm make equal contributions to the fertilized egg and predicted that the number of chromosomes would be halved during the formation of egg and sperm to prevent an increase in chromosome number from generation to generation.

The details of this "reduction division" were discovered in 1902 by W.S. Sutton,[1] then a graduate student at Columbia University. He showed that during meiotic cell divisions a chromosome originally derived from the organism's father pairs with a chromosome of identical size and shape derived from the mother. He showed that these paired homologous chromosomes separate into different germ cells, as would be required if they carry Mendel's paired determinants. He also showed that it is a matter of chance which member of each pair goes into which cell. As a result, Sutton proposed that "the association of paternal and maternal chromosomes in pairs and their subsequent separation during the reducing division ... may constitute the physical basis of the Mendelian law of heredity."

A flood of experiments followed, and in a few years character differences in over 100 plants and 100 animals had been found to be inherited like those in Mendel's peas. The most important work was done by T.H. Morgan and his colleagues at Columbia University.[1] He chose to study the fruit fly *Drosophila melanogaster*. He bred large numbers rapidly in milk bottles and within a few months had demonstrated the mendelian inheritance of many unit characters. As Sutton had pointed out, the number of unit characters in an organism must greatly exceed the number of chromosomes. Hence, if Mendel's determinants (genes) lie on chromosomes, they could not all move independently as found by Mendel but should fall into **linkage groups** equal in number to the number of chromosomes. Morgan proved that *Drosophila melanogaster*, with four pairs of chromosomes, has four pairs of linkage groups and that the relative number of genes in each group is roughly the same as the relative sizes of the four chromosomes: two other species of *Drosophila* with five and six pairs of chromosomes were found to have five and six pairs of linkage groups.

Morgan's study[1] of linkage also resulted in an important step toward discovering the molecular structure of genes by showing that they are arranged one after another along the chromosome. Morgan found that linkage between genes was sometimes broken. For example, if a female fly had a certain two dominant genes in one chromosome and the two recessive alleles in the homologous chromosome, a proportion of its offspring would have one dominant and one recessive gene in each chromosome, as shown by examination of the characteristics of their progeny. This **recombination** was later proved to be due to breakage—during meiotic pairing—of homologous chromosomes at the same position along their length and the rejoining of the fragments from different members of the pair. Different pairs of genes on a chromosome were found to recombine with different frequencies, and examination of these frequencies suggested that genes are arranged along a chromosome like beads threaded along a string.

Later studies in bacteria and viruses, as well as in *Drosophila*, showed that recombination can occur just as readily within a gene as between genes, suggesting that there is no abrupt change in molecular structure between one gene and the next. By this time, it had been shown that DNA is largely concentrated in the chromosomes and that uptake of pure DNA can transfer a gene from one bacterium or virus to another. It was concluded from this and many supporting experiments that each gene is a portion of a very long DNA molecule, which extends the length of the chromosome. Further experiments proved that the molecular structure of a particular gene results in the formation of a protein of a particular molecular structure, a process described as **one gene–one protein** when first discovered. These proteins determine the molecular reactions of the cell, and, as a result, DNA controls the formation or entry into the cell of every other cell molecule.

Hard work by many molecular biologists led to the discovery of the molecular structure of DNA and proteins. They also discovered the chemical reactions by which DNA forms self-copies and by which it directs the formation of proteins. Both DNA and protein molecules are long chains formed by the covalent linking of smaller molecules. Four kinds of closely related **nucleotides** make up DNA molecules (known as A, T, G, and C after the first letters of their chemical names). Twenty kinds of closely related **amino acids** make up proteins. Differences between DNA molecules and between protein molecules depend on the order in which the four different letters are arranged and the total number in the chain.) The particular conformation (three-dimensional) in which DNA molecules exist in chromosomes was discovered in 1953 by James Watson and Francis Crick.[4,5] They occur in complementary pairs of equal length, held side by side by weak noncovalent bonds between the nucleotides, with the "beginning" of one molecule adjacent to the "end" of its partner (**Fig. 5-1**). (The two ends of a DNA molecule can be distinguished because the nucleotides of which it is composed are not symmetric. The nucleotide at one end—by convention named the beginning or "3' end" has an uncombined 3'-OH group; that at the other 5' end has an uncombined 5'-phosphate group.) The complementary pair of molecules is usually known as "a" DNA molecule, and the members of the pair as "strands," although this is not strict chemical terminology. Wherever A occurs in one molecule, T is adjacent in the paired molecule, for the reason that only these two nucleotides lie the correct distance apart to be stabilized by noncovalent bonds. Similarly, wherever G occurs in one molecule, C is adjacent in the paired molecule. This was the greatest discovery in cell biology ever made, since it revealed the molecular foundation of the self-copying of living cells and effectively ruled out vitalist explanations of cell function. It correctly implied that before cell division the paired DNA molecules separate, and, as they do so, a new partner is correctly built up beside each molecule by A pairing with T and by G pairing with C. In this way, each chromosome, by directing its self-copying by enzymes, gives rise to two identical chromosomes that pass into different daughter cells. Watson and Crick also proved that the paired DNA molecules are elegantly coiled into a helix or corkscrew, a

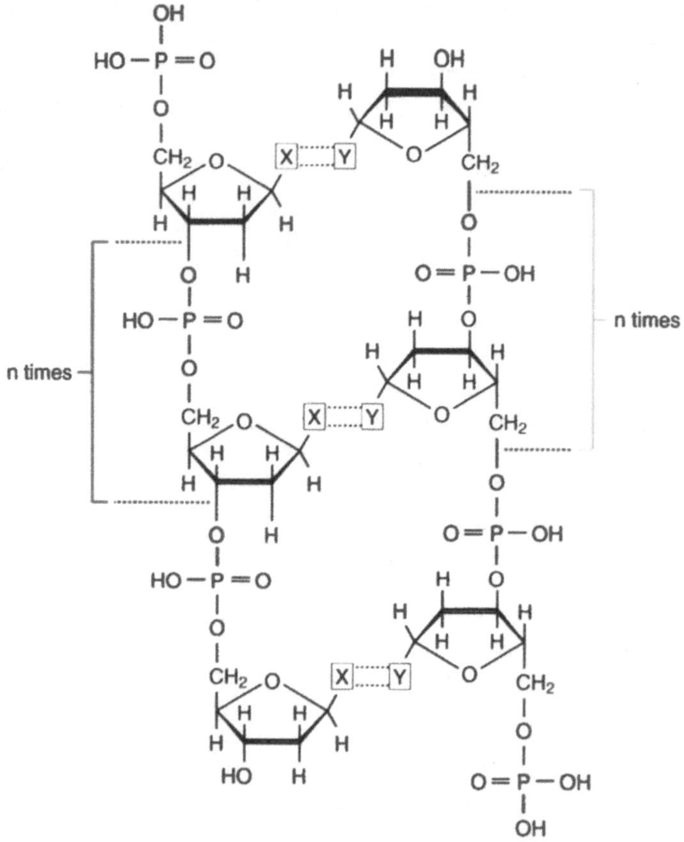

FIGURE 5-1 The chemical structure of two DNA molecules bound together as suggested by Watson and Crick. X and Y represent the bases that are linked by hydrogen bonds. When X is adenine, Y is thymine, and vice versa. When X is guanine, Y is cytosine, and vice versa.

shape assumed by many long molecules and by long structures under inner tension, such as wood or metal shavings. Although this double helix has had great notoriety, it is not immediately relevant to understanding the molecular basis of cell replication.

One of the complementary DNA strands of the gene (the **coding strand**) directs the formation of RNA with the help of the enzyme RNA polymerase, a process known as **transcription** of DNA into RNA. RNA and DNA molecules have a similar structure except that the four nucleotides of RNA are slightly different and are known as A, U, G, and C after the first letters of their chemical names. When a DNA molecule is transcribed, the process is similar to that in the replication of DNA. Wherever A, G, C, or T occur along a DNA strand, then U, C, G, or A, respectively, will be inserted along the RNA molecule. This RNA is then converted to **messenger RNA**: certain nucleotide sequences are excised by enzymes, and some small groups of atoms are added. The second stage is the **translation** of the messenger RNA into protein. Each particular sequence of three nucleotides along a messenger RNA molecule, by a series of molecular reactions catalyzed by enzymes, directs one particular amino acid into the protein chain. In this way, the "spelling" of the messenger RNA is translated into the "spelling" of the protein. Since the transcription of DNA into RNA and the translation of messenger into protein are like the deciphering of one code into another, the DNA is said to **code** for the RNA and the RNA to code for the protein.

The discovery that the DNA of genes directs the formation of proteins is of fundamental importance in development. But as Ross Harrison complained in 1937, "Already we have theories that refer the processes of development to gene action, and regard the whole performance as no more than the realisation of the potencies of genes."[6] Genes are often said to contain information ("programs") from which the structure of the whole adult organism could be deduced, as Schrödinger claimed in 1944, *What is Life*, "In calling the structure of the chromosome fibres a code-script we mean that the all-penetrating mind, once conceived by Laplace ... could tell from their structure whether the egg would develop, under suitable conditions, into a black cock or into a speckled hen, into a fly or a maize plant. ..."[7] In fact, the all-penetrating mind would also at least need to know the structure and three-dimensional distribution of many other molecules within the egg to deduce the patterns of cell division and gene activation, since the program for forming the adult includes all these. Nevertheless, statements still persist that regard the whole process of development as no more than the "realisation of the potencies of genes," as in the following sentences by distinguished scientists: "How the linear information in DNA can generate a three-dimensional organism in the course of development of the fertilised egg is one of the great mysteries of biology" and "how does a 1D gene give a 3D embryo?"

The close interdependence of DNA and other components of the egg is shown by experiments in nuclear transfer. If nuclei from early embryos of *Xenopus laevis* are injected into enucleated eggs of the same species, a high proportion of the eggs develop into adult frogs. But if the nuclei are from the closely related *Xenopus tropicalis*, the eggs never develop beyond an early embryonic stage. Presumably, certain proteins in the egg of *Xenopus laevis* do not have the correct structure or distribution to replicate and activate the *Xenopus tropicalis* genes in a way that will produce an adult *Xenopus tropicalis*. Hence, in the present advanced state of evolution, it appears that to function correctly the genes of an organism require the preexistence of many of the proteins whose formation they direct. How this situation first arose is a matter of conjecture. But the composition and three-dimensional structure of organisms of the present day must have been built up gradually over millions of years by small changes in DNA structure, producing small changes in protein structure that were compatible with correct gene activation from egg to adult. These small changes in turn allowed further changes in DNA sequence to give rise to organisms whose nuclei would not be able to function in the enucleated eggs of their near ancestors. That small changes in protein structure can be compatible with correct gene activation is again shown by nuclear transplantation: if an enucleated egg of *Xenopus laevis* receives a nucleus from an embryo of the subspecies *Xenopus laevis victorianus*, it can develop into an adult *Xenopus laevis victorianus*.[8]

WEISMANN WRONGLY SUGGESTED THAT DIFFERENT CELLS OF AN EMBRYO CONTAIN DIFFERENT GENES

Since genes direct the composition of cells, differences in composition might be expected to arise between embryonic cells because they have been allotted different genes. This was another important suggestion, made in 1892 by August Weismann.[1] Weismann was born in Frankfurt, Germany, and was trained in medicine. After practicing medicine for a short time, he turned to experimental zoology. However, failing eyesight forced him to become a biological theorist. He was outstandingly successful at this, and he greatly accelerated the rate of biological discovery over the next 50 years.

Unlike theoretical physics, of which there are thriving departments in many universities, theoretical biology has never become firmly established. Ever since 1800, enthusiasm for biological theorizing has alternately risen and then fallen in reaction to previous excesses. In the 1940s, it fell into particular disrepute after the collapse of two speculative bubbles of the 1930s: the Bergmann and Niemann theory of protein structure and the tetranucleotide theory of DNA structure. These theories proposed that proteins and DNA are composed of simple repeating sequences of amino acids and nucleotides. It became clear that both theories were based on inadequate analytical measurements and, moreover, that their simplicity would exclude DNA and proteins from the complex control of living cells. There followed a period in which "noses to the experimental grindstone" was the general fashion. This ended dramatically with the announcement of the three-dimensional conformation of DNA by Watson and Crick,[9] which was a triumph of theorizing from the experimental results of others.

Francis Crick went on again to prove that careful theorizing, constantly checked by discussion with colleagues, can significantly increase the rate of biological discovery, as shown by his contributions to our understanding of the mechanism of protein synthesis and the structure of chromatin. Weismann was a similar careful and successful theorist and greatly influenced the direction and success of biological experimentation over 50 years. He was primarily interested in animal and plant development, and, as early as 1892, he suggested that only germ cells retain the full complement of genetic material and that differences between body cells result from their being allotted different portions of this material. In this instance, Weismann was wrong, as he later disarmingly admitted:

> There were two alternatives to explain differentiation: (1) the hypothesis of a systematic and progressive dissection of the genetic potential ... or (2) the hypothesis that the determinants of all characters remain together in all the cells of developing organisms but that each of them is tuned to respond to a specific stimulus that only activates this trait: a pure "dissection" and a pure "activation" theory. I decided in favour of the former, because on the basis of the facts available at that time it seemed to be the more probable one.[1]

Available facts began to move against Weismann's suggestion in the early 1900s as microscopic techniques improved. It became clear that each pair of homologous chromosomes in a cell can be identified by its characteristic structure, and it was seen that different kinds of cell within an organism normally contain a complete chromosome set. This might not exclude the loss of certain genes, but later refinements in microscopy showed that any losses must be restricted to small portions of genes. Thus, if chromosomes of mammalian cells in mitosis are treated with Giemsa stain, they show an irregular series of dark bands characteristic of each chromosome and each containing about 1 million nucleotide pairs from a total of around 1000 million in the chromosomal DNA. There are no visible differences between the bands on homologous chromosomes from different tissues. In addition, **polytene chromosomes** of *Drosophila* (described in the text that follows) have a series of dark bands and light interbands. Each band contains only one or a few genes and about 30,000 base pairs out of around 40 million along the length of a DNA double helix. The banding pattern is identical in several different tissues of *Drosophila*.

A possible variation of Weismann's suggestion that certain genes are lost from certain tissues is that they are activated or inactivated by small changes in nucleotide sequence. For example, a nucleotide sequence near a particular gene might be altered to one that will bind RNA polymerase, thus allowing its transcription into RNA. This change could be brought about by a temporarily formed enzyme, which binds to a unique sequence of nucleotides near the gene. Because the nucleotide sequence of DNA is precisely copied before every cell division, all progeny of this cell would inherit the gene in this active form. For example, the hemoglobin gene of a red blood cell precursor would be converted to an active structure so that all progeny cells would make hemoglobin. Conversely, genes not required by a differentiated cell might be inactivated in a similar way. This mechanism could once again explain the remarkable stability of embryonic cells to changes in environment. To investigate this possibility, many different genes have been isolated from more than one tissue of the same animal, and their nucleotide sequence has been examined, particularly where RNA polymerase binds to initiate transcription. For example, the gene for silk (a protein), together with its neighboring DNA, was isolated from two tissues of the silk moth—the silk gland, where the gene is active, and the pupa, where it is not. No differences in nucleotide sequence were discovered.[10]

Again, some genes in differentiated cells that are not normally active can be artificially activated, thus proving that their DNA is unaltered in structure. For example, human white blood cells can be fused with mouse liver cancer cells to give single cells with two nuclei. On culture, the progeny of such cells form human liver proteins.[11]

That both active and inactive genes of a differentiated cell have essentially the same structure as their ancestral genes in the fertilized egg can be proved by implanting the cell nucleus into an enucleated egg and showing that it can produce a normal adult. This was first done about 40 years ago. Nuclei were taken from frog embryos at the blastula stage and implanted through a fine glass needle into frogs' eggs from which the nuclei had been excised or destroyed by radiation. Well over half the eggs developed into normal tadpoles that gave rise to normal frogs. But when nuclei were taken from older frog embryos, the success rate fell rapidly and was zero with nuclei from adult cells. This at first suggested that genes in differentiated cells have been altered in molecular structure. However, persistent experiments suggested that lack of success came from technical problems, since various pretreatments of the nuclei before implantation gave the following results with low frequency: the formation of normal adult frogs from enucleated eggs implanted with intestinal nuclei of tadpoles and the formation of deformed tadpoles from enucleated eggs implanted with nuclei from adult kidney, heart, lung, skin, and white and red blood cells.

The scene was transformed in 1997 by production of the healthy lamb "Dolly" from an enucleated sheep's egg into which the nucleus of a cell from an adult sheep had been introduced.[12] Cells were isolated from the mammary gland of a 6-year-old Dorset ewe and grown in culture. The culture medium was then changed to one in which the cells were unable to divide. All cells were arrested in the first part of the cell cycle before DNA replication causes the chromosomes to be duplicated, and this appears to have been the key to success. Eggs were taken from Blackface ewes, and their nuclei were removed. A mammary gland cell was placed in contact with an enucleated egg, and the two were made to fuse into a single cell by electrical pulses. The embryos so formed were allowed to undergo a few cell divisions in culture and were then implanted into Blackface ewes, one of which gave birth to the healthy Dorset lamb.

It may be concluded that during embryonic development genes do not generally become lost from tissues or irreversibly activated or repressed by changes in DNA structure that are inherited at subsequent cell divisions. However, there is an important exception. Thousands of different antibody proteins that protect animals against infection are made by white blood cells called **B lymphocytes**. The thousands of different nucleotide sequences that are needed to form these proteins result in part from genetic DNA being severed at different points by enzymes and recombined in different sequences in different cells.

GENES ARE DIFFERENTIALLY ACTIVATED DURING DEVELOPMENT

Since differences in composition between embryonic cells do not result from the loss of different genes, it is necessary to return to the possibility that Weismann at first rejected, namely, "that the determinants of all characters remain together in all the cells of developing organisms but that each of them is tuned to respond to a specific stimulus that only activates this trait."[1] Nowadays, a gene is said to be active when it is being transcribed into RNA that is usually translated into protein. Good evidence exists that differences in embryonic cell composition largely originate from the activation of different genes in different cells and that translation of the resulting RNA into protein is usually rapid. Although translation sometimes awaits a further signal, only a few instances have been found in which genes that are not required by a cell or its progeny are transcribed into RNA.

Clear evidence for consistent patterns of gene activation during a certain stage of development, which are followed by the rapid synthesis of proteins, comes from studies of the "puffing" of polytene chromosomes of the insect larvae, of *Drosophila* and *Chironomus*.[13] Polytene chromosomes are unusual and occur in certain cells of larvae. Rather than the usual single DNA double helix extending the length of the chromosome, many double helixes are arranged side by side with identical genes adjacent. Unlike normal chromosomes, polytenes are visible in a microscope even when cell division is not taking place. They are seen to have an alternating series of dark bands of differing sizes and lighter interbands. Each band usually contains only one gene, and its length of DNA is repeated, side by side, up to 1024 times (i.e., 2^{10} from 10 doublings of the quantity of DNA). At certain stages of development, certain bands swell to form **puffs**, which later regress. A puff has been proved to be the site of RNA formation and hence is a visible manifestation of an active gene.

The formation and regression of puffs in *Drosophila* larvae follow the same sequence during the development of every individual. In the salivary gland chromosomes, about 135 bands puff over 12 hours and later regress, each at precise and characteristic times. At the same time, polytene chromosomes in other tissues undergo different sequences of puffing. The puffing of certain bands is associated with the immediate appearance of certain proteins. For instance, when larvae are warmed to 37° C, six bands in salivary gland chromosomes rapidly puff, and protective **heat shock proteins** soon accumulate. Also, in *Chironomus* the puffing of a certain band is associated with the synthesis of a salivary protein. A related insect has no puffs on this band, and the salivary protein is absent. If the two species are crossed, the progeny have a puff only on the chromosome derived from the *Chironomus* parent, and they have an intermediate amount of the protein.

Other facts show that changes in embryonic cell composition can be initiated by activation of particular genes to form RNA, which is soon translated into the corresponding protein. When a protein is absent from a cell, the corresponding gene is usually inactive; when it is present, the gene is usually being actively transcribed into RNA. For example, red blood cells contain the protein hemoglobin. In chick embryos, cells that are the precursors of red blood cells contain no detectable messenger RNA for hemoglobin until formation of this protein begins, when the RNA rises in quantity by at least 10^5 times. This might be explained by hemoglobin genes being continually transcribed into RNA that is rapidly destroyed unless the protein is being formed, but this was disproved as follows.

Nuclei were isolated from the same precursor cells and incubated with the nucleotides used for RNA formation, one of which was labeled with radioactive hydrogen. Under these conditions, nucleotides are incorporated only into RNA whose transcription had already begun in the intact cell. The total RNA was isolated, and the amount transcribed from the hemoglobin gene was discovered by finding the amount of radioactivity that would hybridize to a sample of the DNA of the gene. Nuclei from cells that had begun to make hemoglobin protein contained at least 100 times as much RNA transcribed from the hemoglobin gene as nuclei from cells that had not.[14]

It must be concluded that many (and probably most) changes in cell composition during development originate from the transcription of particular genes into RNA, which is soon converted to messenger RNA and translated into protein. However, the messenger RNA sometimes lies dormant until a further signal initiates synthesis of the protein, which can then be very rapid because it does not have to await transcription of the gene. Important instances of delayed translation of messenger RNA into protein occur in eggs before and immediately after fertilization.

Many years ago, eggs of sea urchins and frogs that have been enucleated and hence cannot transcribe genes, were shown to divide to give rise to blastulae, with accompanying synthesis of the correct proteins. That this is due to the translation of stored messenger RNAs was suggested by experiments in which drugs that specifically inhibit either transcription of genes or translation of RNA were added to newly fertilized eggs of sea urchins. When translation was inhibited, the embryos did not develop. But when transcription was inhibited, the embryos developed into normal blastulae whereupon development stopped, suggesting that transcription of new genes was now essential.[15] Stored messenger RNAs appear to be present in the eggs of most animals and those for many named proteins have been identified in sea urchin eggs. They are translated at different times before and after fertilization and enable the rapid production of proteins needed for the early cell divisions. How the dormant messenger RNAs are activated is still not fully understood.

When a gene is activated to give messenger RNA, long nucleotide sequences of DNA that do not appear in the final messenger are usually transcribed. These sequences must be removed and degraded within the nucleus before the messenger RNA passes through pores in the nucleus to be translated into protein in the cytoplasm. It has been found that in early embryos of sea urchins this degradation extends—in certain tissues and not in others—to the whole transcripts of certain genes with the result that the corresponding proteins are not formed in the cytoplasm. Hence, in these instances differences in tissue composition result not from differential activation of genes into RNA but from differential destruction of their RNA transcripts. In sea urchins, this process is restricted to early embryos and appears to occur in some other early embryos.

TWO-STEP ACTIVATION OF EMBRYONIC GENES

Activation of genes in response to signal molecules occurs in bacteria as well as in higher organisms. For example, E. coli bacteria normally make the enzyme β-galactosidase, which hydrolyzes lactose to glucose and galactose when surrounded by a solution of lactose. Hence, in the human gut these bacteria do not waste energy making the enzyme unless their host has recently provided them with milk sugar as a nutrient. The mechanism by which the bacteria do this was elucidated in the 1940s in France by Jacques Monod and Francois Jacob, and it is simple. The inactive gene for β-galactosidase has a repressor protein bound to it, which prevents the access of RNA polymerase. Lactose is converted in the cell to a closely related compound that binds to the repressor protein, thus changing its three-dimensional conformation with loss of affinity for the gene and allowing the access of RNA polymerase. If lactose is removed from the cell culture, the process is immediately reversed and the gene again becomes inactive.

Gene activation in embryonic cells has a feature that shows that the mechanism is more complex: although most changes in cell composition during development are temporary, they remain stable in the absence of the signal that induced them, unlike changes induced in the composition of bacterial cells. These facts were implied by Spemann's early experiments (see Chapter 4). He transplanted epidermis in a newt's gastrula from the upper to the lower side, and the transplants later formed islands of nerve tissue within the recipient epidermis. The transplanted cells must already have undergone invisible changes in composition that directed them to form nerve cells, and these changes were stable in the foreign surroundings. Such a tissue is traditionally said to have become **determined** toward a particular fate, but the term is difficult to define precisely and can cause unnecessary anguish to readers who may not realize this.

Dramatic demonstrations of such stable and hidden differences in the composition of developing cells have come from studies of *Drosophila*. Remember that the adult fly largely arises from various imaginal discs that lie dormant within the larva and are all composed of cells of similar appearance (see Chapter 1). If a piece of an imaginal disc is excised from one larva and implanted into the body cavity of another, it forms the expected body structures when the hormone **ecdysone** transforms the larva into a fly: for example, a fragment of a leg disc might form claws.

A piece of an imaginal disc can also be implanted into the body cavity of an adult fly, where it does not encounter the hormone and so undergoes cell division without changing its structure. Adult flies live for only a few weeks, but part of the implant can be put into another fly and then into another. In this way, the cells of the implanted tissue have been kept dividing for many years without change in appearance. Nevertheless, if transferred into the abdomen of a larva, they still form adult tissues when the larva receives the signal from the hormone ecdysone to become a fly, and they almost always form the tissue expected of the original disc.

Cells from mammalian embryos and adults also show similar stability in a foreign environment. Epidermal cells have been isolated from the tail of a rat and injected into the kidney of another rat. Here, they pursued their normal development to form mounds of skin, with hairs and sebaceous glands. Again, pigment cells can be isolated from the retinas of chick embryos and cultured on layers of nutrient gel. Single cells have been taken from such cultures and further subcultured to give over 50 divisions from the original cells. Throughout, although in a foreign environment, the cells retained their normal shape and pigmentation.

About 1970, a discovery was made that appears to provide an important clue toward understanding the stable commitment of developing cells to a future fate.[16] Experiments suggested that activation of a particular gene occurs in two steps, which can take place in different generations of cells: the gene is first made accessible by a stable structural change that can be inherited by progeny cells, and transcription of the gene into RNA results from the access of proteins formed later. The first experiments to suggest this two-step process arose as follows.

When RNA polymerase from *E. coli* was added to nuclei isolated from red blood cells, the gene for hemoglobin—the main protein of these cells—was transcribed into RNA. But when the polymerase was added to nuclei from liver or brain cells, the hemoglobin gene was not transcribed. This suggested that in red blood cells, not in other cells, hemoglobin genes had become accessible to the small molecules of this enzyme by some change in adhering proteins. Accordingly, the accessibility of the gene to the enzyme DNase 1 was tested. This can be isolated from pancreas, and its normal function is to digest DNA of food within the gut. Nuclei from red blood cells of chick embryos were briefly incubated with the enzyme so that only 10% of their DNA was digested. Although only 10% of the genes for the egg protein ovalbumin had been digested, the genes for hemoglobin had totally disappeared. When nuclei from brain cells of chick embryos were similarly digested, the hemoglobin genes were not preferentially removed. Similar experiments were then done on nuclei from hen oviducts in which about 70% of the cells make ovalbumin. When briefly incubated with DNase 1 to digest only 10% of the total DNA, about 70% of the ovalbumin genes but only 10% of the hemoglobin genes were removed.

It became clear that increased sensitivity of genes to DNase is not necessarily associated with their immediate activity. The sensitivity of active genes other than ovalbumin in nuclei from hen oviducts was studied. Because these genes make only small amounts of protein, they are rarely transcribed. Nevertheless, their DNA was also found to be preferentially digested by DNase.[17] In addition, the hemoglobin gene was found to be still accessible to digestion in nuclei from red blood cells of adult hens in which the gene has become inactive. Also, in red blood cells of chick embryos that transcribe only the gene for the embryonic form of hemoglobin, the genes for both embryonic and adult hemoglobin were found to be equally sensitive to DNase digestion. It was suggested that "the embryonic red cell has in some way recognized and marked the adult globin gene before it is actually expressed. ..."[18] It was later discovered that within these long sequences of nucleotides in DNA that are digested by DNase 1 may be short "hypersensitive sites," which are digested by low concentrations of DNase 1 and leave the remaining nucleotides untouched. These appear to provide access of proteins, such as transcription factors, to sites of gene regulation.

Hence, when embryonic cells become committed to a certain fate, local self-sustaining changes in chromosome structure may be induced, which allow the later access of proteins to genes whose transcription is needed to fulfil that fate. These ideas are supported by studies of gene activity in cultures of **L-cells**, a strain of cancer cell derived from mice. If fragments of chromosomal DNA from other cells are added to the culture medium under the correct conditions, some L-cells will take up a fragment and incorporate it into the DNA of one of their chromosomes, where it may direct the formation of a protein. Pure cultures can then be made from single cells and each culture tested for the presence of a desired extra gene. In this way, a culture of L-cells was obtained that had incorporated the gene for ovalbumin from mice. The cells were found to make ovalbumin protein, although the original L-cells did not. Hence, these cells contained the two normal copies of the mouse ovalbumin gene that were inactive, plus an extra one that was active, and this situation was transmitted to their progeny cells. This proves that even though a cell may contain all diffusible molecules needed for the transcription of a certain gene into messenger RNA, the gene can be maintained over several cell generations in an inactive state.

These findings strongly suggest that during development many genes are activated in two steps.[19] Chromosomes of higher organisms, unlike those of bacteria, have many kinds of structural proteins attached to their DNA by weak chemical bonds. The double-helical DNA molecule that runs the length of the chromosome is coiled into a further helix (i.e., corkscrew) around particles called **nucleosomes**, which are composed of four **histone** proteins. This coiled-coil is further compacted into loops by a fifth histone, H1. It is probable that the first step in gene activation involves a local decrease in the binding of H1.

SPECIFIC TRANSCRIPTION FACTORS ARE KEYS TO THE FINAL STEP IN GENE ACTIVATION

In *E. coli* bacteria, unless transcription is blocked by a repressor protein, most genes are continuously transcribed by the single RNA polymerase of the cell, and the messenger RNA molecules so formed are immediately translated into proteins. In higher organisms, the mechanism of gene transcription is more complex.[20] There are three RNA polymerases, and genes that code for proteins are transcribed by RNA polymerase II. Genes do not lie immediately adjacent to one another along the chromosomes but are separated by long stretches of DNA whose function is in part unknown. Moreover, the sequence of nucleotides in a gene that codes for the sequence of amino acids in the protein is usually interrupted by one or a few **introns**, sequences of noncoding nucleotides. RNA polymerase II transcribes the coding regions, the introns, and some of the noncoding DNA on either side. After transcription, the RNA must be processed within the nucleus to give messenger RNA: the transcripts of the introns are excised and nucleotides are removed from either end and certain others are added. (The nucleotides that are excised occasionally differ from one type of cell to another with the result that proteins with somewhat different amino acid sequences are formed.) The resulting messenger RNA is not translated into protein molecules until it passes through nuclear pores into the cytoplasm.

Moreover, in higher organisms, genes are not transcribed by RNA polymerase II unless other proteins called **transcription factors** are present.[21] Some of these are required by nearly all genes, but other "specific" transcription factors are required by only one or a few genes. Discoveries over recent years have revealed that specific transcription factors are the keys that unlock different genes within different cells during embryonic development. Studies of B lymphocytes first suggested that specific transcription factors play a key role in animal development. They showed that a certain factor can be peculiar to one type of differentiated cell and can be responsible for gene activity unique to that cell. B lymphocytes circulate in the blood and secrete antibody proteins (immunoglobulins), which protect the animal against infection. Two genes code for two proteins (short **light chains** and long **heavy chains**), and two molecules of each kind associate to form an immunoglobulin molecule. These genes are transcribed into RNA only in B lymphocytes. The genes and their adjacent DNA have been isolated from several species of mammal and their structure determined. In every species, the sequence of eight nucleotides ATTTGCAT on one of the two complementary DNA strands of the double helix is found about 70 base pairs before the start of each gene and is not found before other genes.

As previously mentioned, if cells are cultured with DNA in their medium, some cells take up a DNA fragment and incorporate it into a chromosome. In this way, cultures of lymphocytes, and some other types of cell, which had incorporated one of these two genes plus adjacent DNA from a different species were obtained. Provided that this DNA had been incorporated into lymphocyte cells and that the sequence of eight nucleotides had not been removed from the adjacent DNA, the genes were transcribed into RNA as shown by its hybridization to a radioactive copy of the gene from the different species. Otherwise, they were not transcribed.

These facts suggested that the immune protein genes are activated when a protein transcription factor that is peculiar to lymphocytes binds to the eight nucleotides. This was confirmed by covalently linking DNA, which contained the sequence of eight nucleotides to the gene for hemoglobin that is not normally transcribed in lymphocytes. When this modified gene was incorporated into the chromosomes of lymphocyte cells (but not when incorporated into those of other cells), the hemoglobin gene was transcribed into RNA. It was possible to isolate the specific transcription factor by searching for a protein peculiar to lymphocytes that bound to the sequence of eight nucleotides. This protein was called **Oct-2**.

Since that time, many other specific transcription factors peculiar to a single type of cell have been characterized, and the nucleotide sequences to which they bind have been identified. In fact, this knowledge is put to use in the commercial production of certain proteins. Proteins involved in blood clotting, which are lacking in people with hemophilia, are produced in this way. DNA that codes for the blood clotting protein is joined to DNA adjacent to a gene for one of the proteins of sheep's milk, which binds specific transcription factors unique to the mammary gland. This DNA combination is injected into the nucleus of a fertilized egg of a sheep, which is then implanted in the uterus of a foster mother. The resulting female lambs are reared to give rise to ewes. If the injected DNA has been incorporated into a chromosome, the milk from some of the ewes contains the blood-clotting protein that can be isolated.

Many facts are now known about gene transcription in animals and about the transcription factors that activate it. The structure of a typical gene that codes for a protein, and its adjacent DNA, is shown in **Figure 5-2**. Centered about 30 base pairs before the start of the sequence that is transcribed into

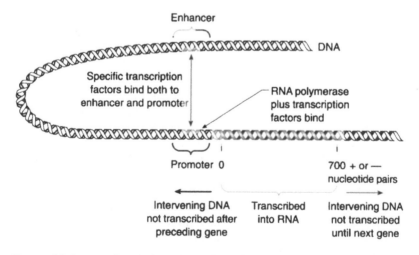

FIGURE 5-2 Structure of a typical gene that codes for a protein, together with its adjacent DNA.

RNA is the **promoter**, a sequence of nucleotides that binds RNA polymerase II and also certain transcription factors needed for its action, some of which (e.g., general transcription factors) are common to many genes that code for proteins. The coding sequence is usually interrupted by one or more introns. These and other nucleotide sequences—known as **enhancers**—in the long intervening tracts of DNA between genes can also bind specific transcription factors. Enhancers can occur in up to thousands of nucleotides before or after the gene, and transcription factors that bind to distant enhancers can interact with the promoter and influence the rate of gene transcription through looping of the DNA.

Some specific transcription factors merely activate the transcription of one particular gene, but others may do the following:

- Activate the transcription of several genes.
- Repress the transcription of one or more genes.
- Activate certain genes and repress certain others.
- Activate a gene when there are few molecules of the transcription factor within a cell, and repress it when there are many (or vice versa).
- Bind to another transcription factor to produce effects not achieved by either alone.
- Activate the gene that directs its own formation, thus ensuring that, once initiated by some other molecule, formation of the transcription factor will continue.

As a result, the formation or activation within a cell of a specific transcription factor can produce one of the many subtle changes in gene activity that development from egg to adult requires.

Like most other proteins, transcription factors are coiled into a three-dimensional structure (**conformation**) whose shape is determined by their particular sequence of amino acids. Some of the component amino acids are brought together by this coiling into exactly the arrangement required to form weak bonds with certain nucleotide sequences in DNA or with molecules involved in gene regulation. A sequence of up to 100 amino acids is coiled into a structure that binds to DNA, and variations in the sequence cause different transcription factors to bind to different nucleotide sequences. Other stretches of 30 to 100 amino acids within the transcription factor bind to protein and other molecules that influence the activity of the adjacent gene. Clearly, animals can rapidly evolve transcription factors with different characteristics because the regulatory amino acids are separated into these different modules: mutations in genes that code for transcription factors can subtly alter the properties of one module without affecting another.

The amino acids that bind to DNA are largely arranged in one of three types of conformation: **zinc fingers** (hence a dietary requirement for zinc), **leucine zippers**, and **homeodomains**. The nucleotide sequence in a gene that codes for the amino acid sequence of a homeodomain within a transcription factor is called a **homeobox**. All homeoboxes are closely related in sequence and hence can be detected in any sample of DNA by its hybridization to DNA known to contain a homeobox. In this way, they have been found in a wide range of animals, suggesting that this class of transcription factor is involved in development throughout the animal kingdom.

TECHNIQUES FOR STUDYING THE ACTION OF GENES IN DEVELOPMENT

Many discoveries concerning the molecular basis of development are founded on some remarkable, but basically simple, laboratory techniques for manipulating nucleic acids that will be outlined here. The reader must understand the word "pure" as used by chemists. A **pure** sample of a chemical compound is one in which all molecules are identical, that is, each molecule contains the same number of atoms of each kind linked together in the same way.

1. *Nucleic acid hybridization.* When a solution of double-helical DNA is warmed to around 80° C, the noncovalent bonds that bind the complementary molecules together are broken, and they separate as single strands that fold into random coils (**Fig. 5-3**). This **denaturation** can be easily followed by putting the DNA solution in the glass cell of a photometer: as the complementary molecules separate (also known confusingly as "melting"), the absorption of ultraviolet light by the solution rises to a maximum. If the temperature of the solution is then lowered to about 55° C, the light absorption returns to its original value as the complementary molecules reform the double helix, or **reanneal**. If an excess of RNA with a base sequence that is complementary to one of these paired DNA molecules (such as RNA that would be formed by transcribing the DNA with RNA polymerase) is added to the denatured DNA solution, then the RNA will preferentially **hybridize**, on cooling, with its complementary DNA. Almost all laboratory techniques for studying gene action at some stage involve DNA:RNA hybridization or the reannealing of DNA to DNA.

2. *Use of enzymes that act on DNA.* Nearly all techniques for manipulating DNA make use of enzymes that bacteria and some other living organisms have evolved for synthesizing, breaking, or rejoining DNA molecules.

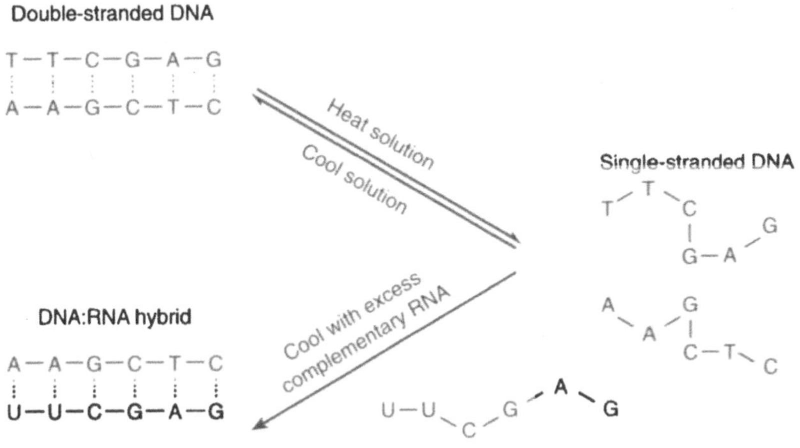

Double-stranded DNA

T — T — C — G — A — G
⋮   ⋮   ⋮   ⋮   ⋮   ⋮
A — A — G — C — T — C

Heat solution
Cool solution

Single-stranded DNA

T ⌐ T ⌐ C
           │   G
        G — A ⌐

A ⌐   G
    A ⌐ │
        C — T ⌐ C

DNA:RNA hybrid

A — A — G — C — T — C
⋮   ⋮   ⋮   ⋮   ⋮   ⋮
U — U — C — G — A — G

Cool with excess complementary RNA

U — U ⌐   G ⌐ A ⌐ G
        C

FIGURE 5-3 Simple example of "melting" and "reannealing" of DNA and of hybridization of one DNA strand to complementary RNA.

Nucleases sever covalent linkages between nucleotides: **endonucleases** act within a molecule,whereas **exonucleases** remove nucleotides one at a time from one of the ends of a molecule. Very important are **restriction endonucleases** (also called **restriction enzymes**), which were discovered in the 1960s. They occur in many bacteria, where they destroy foreign DNA that may invade and harm the cell. The bacteria protect their own DNA by adding small groups of atoms to the nucleotide sequence that the enzyme attacks. Over 400 restriction endonucleases have been isolated, and many can be bought cheaply from chemical supply companies. Each binds to a particular short sequence in double-helical DNA and severs each strand at points that are either adjacent or two, three, or four nucleotides apart, when the two severed ends of the double helix will have protruding short single complementary strands (known as "sticky ends") (**Fig. 5-4**). As a result, the end of one DNA fragment will anneal with the end of the other or with the severed end of any other DNA that has been broken by the same restriction enzyme. This allows the joining of fragments from different DNA molecules by the enzyme **DNA ligase**. Restriction enzymes enable the preparation from very large DNA molecules of pure lengths of DNA of a manageable size for sequencing and other operations.

There is a simple technique for separating DNA molecules according to their length, namely, **gel electrophoresis**. In this process, the mixtures of molecules are placed in a depression near a negative electrode at one end of a flat gel in a rectangular trough and attracted toward a positive electrode at the other end. The molecules move toward the electrode at a speed in

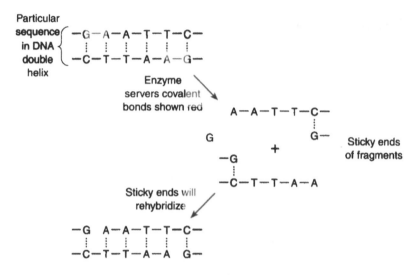

FIGURE 5-4 Action of *Eco*RI, a restriction enzyme from *Escherichia coli.*

inverse proportion to their length, and their positions can be detected by their absorption of ultraviolet light. Other important enzymes are **DNA polymerases**, which build up a DNA molecule that is complementary to another single-stranded DNA molecule from the four component nucleotides of DNA in their active form (namely, the four deoxynucleoside triphosphates), and **reverse transcriptases**, which build up a DNA molecule that is complementary to a molecule of RNA.

3. *Cloning of DNA in bacteria.* Since most chromosomes have only one DNA double helix along their length, the task of discovering the molecular basis of gene action in development would be extremely difficult if there were no mechanism for copying DNA in the laboratory to give workable quantities. Fortunately, methods for doing this have been developed. The process is known as **cloning**: just as a clone of cells is a population of cells formed from a single cell, so a clone of DNA molecules is a population formed from a single molecule. Cloning in bacteria makes use of the fact that *E. coli* cells often contain, in addition to their single, large, circular chromosome of double-helical DNA, a small circular minichromosome, usually in multiple copies, known as a **plasmid**. If a plasmid is opened by a restriction endonuclease that forms sticky ends, these ends anneal with the ends of any linear DNA that has been produced by digestion with the same restriction enzyme. Thus, the two molecules can be joined with DNA ligase to give a larger circular plasmid. If *E. coli* cells are grown in a medium that contains these plasmids, some are taken up by the cells and can be replicated in

unlimited quantities. These cells can then be isolated, then cleaved by the same restriction enzyme, and the cloned DNA can be separated from the original plasmid DNA. This basic technique has been improved in many ways. Artificial plasmids can be bought from chemical supply companies, which have a length of DNA inserted with sites for a number of restriction enzymes, so widening their use. The plasmids also have a gene for antibiotic resistance, thus enabling bacteria that have not incorporated a plasmid to be killed by the antibiotic. In addition, they can include the gene for the enzyme β-galactosidase, which is severed and inactivated when the DNA to be cloned is incorporated. After culture, the bacteria are grown on nutrient gel to give rise to colonies derived from single cells. Only colonies that do not give a color reaction for β-galactosidase, and hence have incorporated the gene to be cloned, are selected for further growth. Also, other "vectors" can be bought for cloning DNA molecules that are larger than those that can be cloned in plasmids.

4. *Polymerase chain reaction.*[22] This ingenious technique for amplifying (i.e., cloning) minute quantities of DNA by chemical reactions directed by DNA polymerase was devised in 1984 by Kary Mullis[22] and caused an immediate transformation of DNA technology for which he was rapidly awarded the Nobel Prize. The method is simpler and quicker than cloning in bacteria, and the DNA can contain as many as 10,000 base pairs, and even a single double helix can be amplified. Three separate steps taking a few minutes bring about the doubling of DNA, and these can be repeated many times, thus causing an exponential increase in quantity (**Fig. 5-5**). The reaction solution contains the double-helical DNA to be amplified, the four nucleoside triphosphates, and a DNA polymerase enzyme. It must also contain two "primer" DNAs complementary to about 20 nucleotides at opposite ends of the double helix that is to be amplified. In the *first* step, the solution is heated to 95° C for 15 seconds to separate the complementary molecules of the double helix. In the *second* step, the solution is cooled to 54° C; the primers are in excess, and they rapidly hybridize to the 5′ ends of the separated DNA molecules without their reforming a double helix. In the *third* step, the solution is incubated to allow the polymerase enzyme to catalyze the extension of each primer to give rise to two new double helixes identical with the original. In the Mark 1 version of the polymerase chain reaction, RNA polymerase from *E. coli* was used. This is destroyed by heating to 95° C and hence had to be replaced before the three steps could be repeated. But Mullis[22] had the brilliant idea of using the polymerase of a bacterium that lives in hot springs and hence can withstand the heating. As a result, the three steps can be repeated many times without any additions to the solution, thus giving a geometric increase in the quantity of DNA. After 30 repeats, which can be

STEP 1
Heat to 95° for 15 sec to denature DNA

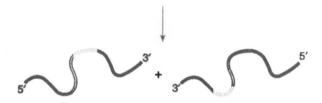

STEP 2
Rapidly cool to 54°. Primers hybridize to ends of each strand.

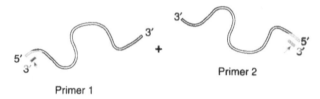

Primer 1

Primer 2

STEP 3
Incubate to allow extension of primers by RNA polymerase.

FIGURE 5-5 The three steps of the polymerase chain reaction.

carried out within 1 hour, the DNA is amplified in quantity by 1000 million times. The gene for the polymerase has been transferred to another bacterium that can be easily grown; this allows the enzyme to be produced in quantity and it is now sold at a low price. In addition, a small inexpensive machine can be bought that will perform the steps of the reaction automatically.

5. *Sequencing DNA*. The most frequently used method for finding the nucleotide sequence of a pure sample of DNA is that devised in the 1970s by the double Nobel Prize winner Fred Sanger[23] (**Fig. 5-6**). It uses DNA

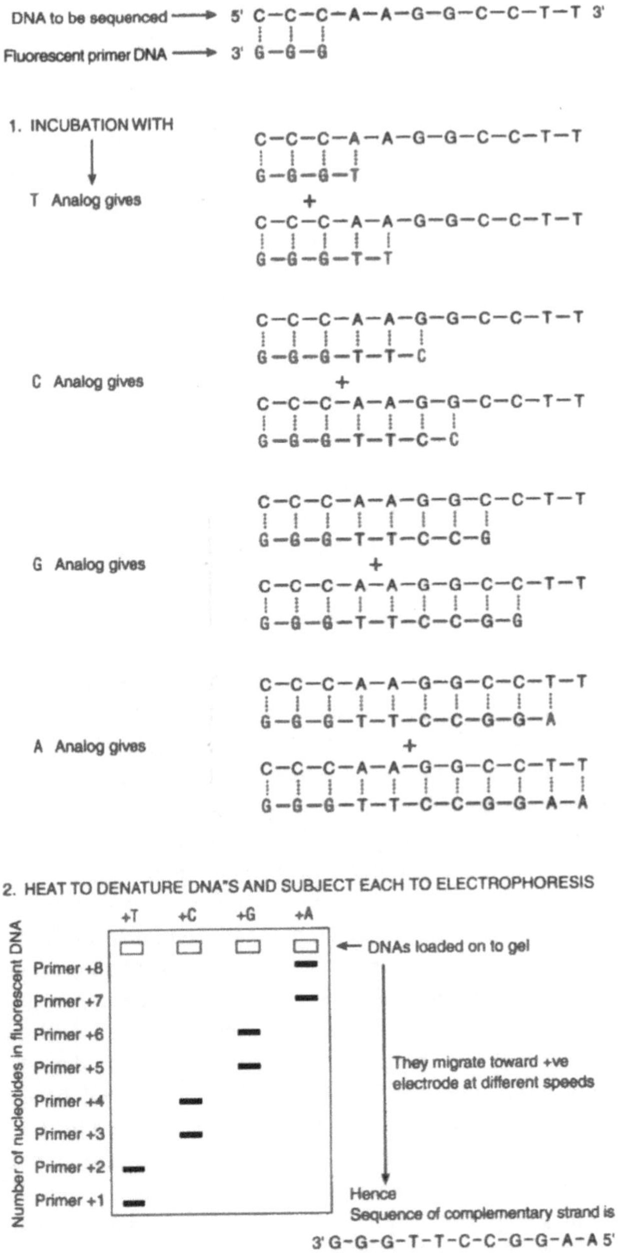
DNA to be sequenced → 5' C—C—C—A—A—G—G—C—C—T—T 3'
Fluorescent primer DNA → 3' G—G—G
1. INCUBATION WITH
T Analog gives
C—C—C—A—A—G—G—C—C—T—T
G—G—G—T
+
C—C—C—A—A—G—G—C—C—T—T
G—G—G—T—T
C Analog gives
C—C—C—A—A—G—G—C—C—T—T
G—G—G—T—T—C
+
C—C—C—A—A—G—G—C—C—T—T
G—G—G—T—T—C—C
G Analog gives
C—C—C—A—A—G—G—C—C—T—T
G—G—G—T—T—C—C—G
+
C—C—C—A—A—G—G—C—C—T—T
G—G—G—T—T—C—C—G—G
A Analog gives
C—C—C—A—A—G—G—C—C—T—T
G—G—G—T—T—C—C—G—G—A
+
C—C—C—A—A—G—G—C—C—T—T
G—G—G—T—T—C—C—G—G—A—A
2. HEAT TO DENATURE DNA'S AND SUBJECT EACH TO ELECTROPHORESIS
+T
+C
+G
+A
DNAs loaded on to gel
Number of nucleotides in fluorescent DNA
Primer +8
Primer +7
Primer +6
Primer +5
Primer +4
Primer +3
Primer +2
Primer +1
They migrate toward +ve electrode at different speeds
Hence
Sequence of complementary strand is
3' G-G-G-T-T-C-C-G-G-A-A 5'

FIGURE 5-6 Simple illustration of Sanger's method for sequencing DNA.

polymerase I from *E. coli* bacteria, where the enzyme is involved in repairing DNA molecules. If this enzyme is added to a solution of the four deoxynucleotide triphosphates, plus the DNA to be sequenced in its single stranded form to which has been hybridized a short primer DNA with a sequence that is complementary to the 5′ end of the molecule, the polymerase will extend the primer to form a full complementary DNA double helix. This exact mixture is used in the sequencing of DNA with one exception: an unnatural analogue of one of the nucleotides (a dideoxynucleoside triphosphate) is added, which is incorporated into the chain but prevents its further extension. Four incubations are set up, one with the analogue for nucleotide A and the others with that for T, G, and C. The concentration of the analogue is adjusted so that sometimes it is inserted instead of the correct nucleotide as the molecule extends, and sometimes it is not. Hence, each of the four incubations contain a number of unfinished molecules of different lengths.

Suppose, for example, that the DNA contains the nucleotide C at 5 and 6 nucleotides from the end of the primer. Then, the incubation with the G analogue will contain molecules of length: primer plus 5 and primer plus 6 nucleotides. The four mixtures are subjected to gel electrophoresis in four adjacent lanes. If the primer contains a fluorescent dye, the molecules of increasing length show up as a series of fluorescent bands at decreasing distances from the origin. In total, the four lanes will contain as many fluorescent bands as there are bases in the newly made DNA. Each band occurs in succession along the length of the gel and in a particular lane, depending on whether its molecule terminates in A, T, G, or C. Hence, the complete sequence of nucleotides can be deduced. From this, the complementary sequence in the sample of DNA can be deduced. Molecules up to 500 nucleotides long can be sequenced in this way, and, with a collection of overlapping fragments prepared with different restriction enzymes, the sequencing can be continued indefinitely.

The Sanger technique, which can now be largely automated, is being used in Britain and the United States to sequence the 3×10^9 pairs of DNA nucleotides that are distributed among the 23 different human chromosomes. It was thought appropriate to combine the DNA from a number of anonymous men and women; hence, there will be variations in sequence in certain regions. Two groups are working on the problem: one in the Human Genome Project in Cambridge England, and another in Celera Genomics, Rockville, Maryland. The Human Genome Project has isolated long fragments whose relative positions along the total DNA are known. Each of these long fragments is broken into a mixture of random fragments that are cloned in bacteria to give pure samples whose nucleotide sequence is determined. The sequences are then fed into a computer that is programmed to detect overlaps by means of the sequence at the end of one fragment being identical with that at the beginning of another. (Just as "seque" and "quence" can be seen to be overlaps in the word "sequence.") In this way, the computer deduces the sequence of each long fragment and hence that of the total DNA. Celera Genomics breaks the total human DNA into small random fragments that are cloned to give pure samples. From overlaps in the sequences of these fragments, the sequence of each of the human chromosomes is gradually built up by computer. This method, although rapid, can meet difficulties in deciphering the repetitive sequences that lie between genes.

In June 2000, the two groups met under the auspices of US President William J. Clinton to announce that a working draft of the nucleotide sequence of human DNA had been achieved, although many regions remain uncertain. Most of the DNA does not code for proteins, but the regions that do so—the genes—can be identified by similarities in sequence. This will enable the total number of genes to be counted: first estimates suggest that the number lies between 38,000 and 115,000. Also, from the nucleotide sequence of a gene, it is usually possible to deduce the amino acid sequence of the corresponding protein and sometimes, from this sequence, the function of the protein. Variations in DNA sequence between different people are likely to be small, but variations that lie within genes can be important because they determine inherited differences among individuals, including inherited differences in health.

6. *Genetic transformation of cells.* If DNA containing a foreign gene is introduced into an animal cell, it can be incorporated into a chromosome and transcribed into RNA; this translates into the protein coded by the gene (this was surprising to most biochemists, who predicted all kinds of pitfalls on the way). The DNA can be introduced into cultured cells by associating it with an RNA virus that acts by incorporating itself into a chromosome. It can also be injected into the male nucleus within a fertilized egg of a mouse from a fine glass needle under a microscope, when it is often incorporated into a chromosome by the cell's enzymes. If the gene has the correct adjacent nucleotide sequence for binding RNA polymerase and transcription factors, it may be transcribed into RNA. This technique can be useful in studying development.

The study of development in mice has recently been revolutionized by the technique of **gene targeting** or **gene knockout**, by which a chosen gene can be modified or inactivated. The procedure is founded on the formation of chimeric mice by injecting cells from the inner cell mass of one early embryo into another similar embryo. The mice that develop after implantation into a foster mother are composed of a mixture of cells from the two embryos: if, for example, one embryo is from a white mouse and the other is from a black mouse, the coat will be a mixture of black and white. (It appears that some, largely unsuspecting, humans are also chimeras formed by the chance fusion of twin embryos. This can be detected if the embryos were of different sexes by some cells of the adult having male sex chromosomes and others female.) For gene targeting, cells are removed from the inner cell mass of a mouse, and their number is increased in culture. The cells are then incubated under conditions in which they absorb added DNA whose structure has been designed to insert itself (by hybridization and enzyme catalysis) into a particular gene of known structure whose activity is thereby eliminated or changed. This occurs in very few of the cells, but they can be selected by killing all others: mouse cells are used that can be killed by an added drug, and cells that incorporate the added DNA are protected by a gene inserted into this DNA, which forms a protective protein. These cells are then injected into normal mouse embryos, and the chimeric mice that result have the altered gene in some of their cells. These mice are then mated with normal mice. Any egg or sperm cells that contain the altered gene give rise to mice that have one altered and one normal gene in each body cell. When these mice are mated with one another, one-fourth of their progeny (as pointed out by Mendel) have two altered genes in each body cell, and the effect of this on development can be studied.

7. *Reverse genetics.* This technique enables the unknown gene that codes for a known protein to be identified and isolated. It can provide useful information when a new protein appears during the development of an animal. The protein is isolated, and part of its amino acid sequence is determined. The nucleotide sequences in the gene that could code for this sequence are deduced, and they, or sequences that are complementary to them, are synthesized with radioactive nucleotides. The total DNA of the animal (e.g., *Drosophila*) is broken into fragments by restriction enzymes, which are incorporated into plasmids that are cloned to give pure samples of each DNA fragment. Any fragments that hybridize with the radioactive DNAs are retained. From the nucleotide sequences of these DNA fragments, the complete nucleotide sequence of the gene and its adjacent regulatory sequences may be deduced.

REFERENCES

1. Mayr E. *The Growth of Biological Thought.* Cambridge MA: Harvard University Press, 1982. (Contains a detailed discussion of the development of our knowledge of inheritance and gene function, with numerous references.)

2. Iltis H. *Life of Mendel*, translated by E. and C. Paul. London: George Allen and Unwin, Ltd, 1932. (A fascinating account of Mendel's life by a scientist from Brunn who spoke to many of his relatives and friends.)

3. Stern C, Sherwood ER, eds. *The Origin of Genetics: A Mendel Source Book*. San Francisco: WH Freeman, 1966. (Contains a translation of Mendel's paper.)

4. Watson JD, Crick FH. Molecular structure of nucleic acids: a structure for deoxyribose nucleic acid. *Nature* 1953;171:737–738.

5. Watson JD, Crick FH. Genetic implications of the structure of deoxyribonucleic acid. *Nature* 1953;171:964–967.

6. Harrison RG. Embryology and its relations. *Science* 1937;85:369–374.

7. Schrödinger E. *What Is Life?* Cambridge, UK: Cambridge University Press 1967:22.

8. Barry JM. Informational DNA: a useful concept? *Trends Biochem Sci* 1986;11: 31–318.

9. Watson JD. *The Double Helix: Being a Personal Account of the Discovery of the Structure of DNA*. New York: Atheneum, 1968.

10. Suzuki Y, Adachi S. Signal sequences associated with fibroin gene expression are identical in fibroin-producer and -nonproducer tissues. *Dev Growth Differ* 1984;26:139–147.

11. DiBernadino MA, Hoffner NJ, Etkin LD. Activation of dormant genes in specialised cells. *Science* 1984;224:946–952.

12. Wilmut I. Cloning for medicine. *Sci Am* 1998;279(No. 6):30–35.

13. Ashburner M. Puffs, genes and hormones revisited. *Cell* 1990;61:1–3.

14. Groudine M, Peretz M, Weintraub H. Transcriptional regulation of hemoglobin switching in chicken embryos. *Mol Cell Biol* 1981;1:281–288.

15. Gagnon MI, Angerer LM, Angerer RC. Posttranscriptional regulation of ectoderm-specific gene expression in early sea urchin embryos. *Development* 1992;114:457–467.

16. Weintraub H, Groudine M. Chromosomal subunits in active genes have an altered conformation. *Science* 1976;193:848–856.

17. Garel A, Zolan M, Axel R. Genes transcribed at diverse rates have a similar conformation in chromatin. *Proc Natl Acad Sci USA* 1977;74:4867–4871.

18. Weintraub H. Assembly and propagation of repressed and derepressed chromosomal states. *Cell* 1985;42:705–711.

19. Orlando V, Paro R. Chromatin multiprotein complexes involved in the maintenance of transcription patterns. *Curr Opin Genet Dev* 1995;5:174–179.

20. Struhl K. Fundamentally different logic of gene regulation in eukaryotes and prokaryotes. *Cell* 1999;98:1–4.

21. Latchman DS. *Eukaryotic Transcription Factors*, 2nd ed. London and New York: Academic Press, 1995.

22. Mullis KB. The unusual origin of the polymerase chain reaction. *Sci Am* 1990;262(No. 4):36–43. (An amusing account of the discovery of the polymerase chain reaction by its discoverer.)

23. Sanger F, Niclen S, Coulson AR. DNA sequencing with chain-terminating inhibitors. *Proc Natl Acad Sci* 1977;74:5463–5467.

CHAPTER SIX

EXPERIMENTS ON *DROSOPHILA* REVEAL THAT SPECIFIC TRANSCRIPTION FACTORS ARE KEYS THAT UNLOCK EMBRYONIC GENES

THE BREAKTHROUGH OF NÜSSLEIN-VOLHARD AND WIESCHAUS

A brilliantly designed program of research was relentlessly pursued in Germany in the 1980s by Christiane Nüsslein-Volhard and Eric Wieschaus.[1] They initiated the discovery in precise molecular detail of how differences in gene activity between the nuclei of the *Drosophila* embryo first arise. For this work, they were awarded the Nobel Prize in 1995. These findings transformed the atmosphere of developmental biology into one of optimism—that the essential features of animal development would soon be understood in molecular terms.

The *Drosophila* egg does not subdivide after fertilization into separate cells. Instead, the nucleus divides, and its progeny subdivide, to give rise to several thousand nuclei lying in the egg cytoplasm. Most of these nuclei migrate to form a single layer under the surface of the egg, where they are later enclosed by membranes to give rise to separate cells. Three conclusions had previously been reached from simple experiments with razor blades, injection needles, and cotton threads (see Chapter 3), about how genes are activated in each nucleus. *First*, differences in gene activity arise from nuclei lying in regions of egg cytoplasm of differing composition. *Second*, communication is needed between the front and rear of the egg to allow these regions to acquire these differences. *Third*, each nucleus receives two separate activating signals, one that is correct for its position along the anterior-posterior axis and another for its position along a line at right angles, namely, the dorsal-ventral axis.

The fact that nuclei in the early *Drosophila* embryo are not separated by cell membranes once made studies of its early development appear somewhat irrelevant to that of other organisms, but it turned out to be a blessing in disguise. Studies of most early embryos are complicated by the fact that intricate signal transductions originating at cell membranes are interposed between signals from one cell nucleus to another. But in *Drosophila*, protein molecules that cannot cross cell membranes can move between nuclei, and this enabled Nüsslein-Volhard and Wieschaus[1,2] to reveal the importance of specific protein transcription factors in inducing differences in gene activity. Unexpectedly, the genes that they first activate do not code for enzymes and structural proteins that give cells their characteristic properties. Rather, they code for other transcription factors that diffuse through the cytoplasm and regulate genes in new and smaller groups of nuclei. These genes again code for further transcription factors, and the process recurs until the nuclei become enclosed in cell membranes. Hence, the newly formed cells foreshadow the body structure of the larva, not by each containing the correct assembly of enzymes and structural proteins but by containing an assortment of specific transcription factors that lead to their formation.

These discoveries were founded on a search for gene mutations in *Drosophila* that alter its early development. **Mutations** are changes in body structure that result from changes in the nucleotide sequence of genes. (The changes in gene structure are also, for convenience, usually called mutations.) In *Drosophila*, mutations can be induced by shining x-rays on flies or by feeding them certain chemicals. A mutation originates in one of a pair of genes that codes for a particular protein, and it either inactivates the gene or causes it to form a protein of abnormal structure. The normal gene on the paired chromosome still makes the normal protein, but it can be replaced by the same mutant gene by cross-breeding flies. Thousands of embryos that were bred from such flies were examined, and mutations in about 140 different genes were found to alter the structure of the early embryo. Clearly, almost all such genes had been discovered, since the same structural change caused by mutation in the same gene, occurred repeatedly.

These developmental mutations are of two kinds. Most are **zygotic mutations** in genes that are active after fertilization and can derive from both the father and the mother. But about 30 are **maternal effect mutations** in genes that act only in the mother to determine the structure of the unfertilized egg. Since there are only about 30 maternal effect mutations and since the changes in embryo structure caused by their mutation are even fewer, differences in gene activity between nuclei can hardly result from the cytoplasm of the unfertilized egg containing a static mosaic of many kinds of inducing molecules, as once suggested. Instead, there must be a simple initial arrangement of a few inducing molecules that increase in number as development proceeds.

Of the maternal effect mutations, 18 altered the embryo structure along the anterior-posterior axis by, for example, deleting certain segments. The other 12 altered the structure along the dorsal-ventral axis by, for example, causing cells typical of the lower regions to encroach upward on either side. This confirmed the suggestion that differences in cell composition along these two directions are largely initiated by separate mechanisms. The meaning of this can be illustrated as follows: the embryo can be idealized as a cylinder with successive rings of nuclei encircling its surface from front to rear, with the same nuclei forming parallel lines along its length (**Fig. 6-1**). Signal molecules acting from front to rear could induce differences in gene expression among nuclei of different rings but would treat nuclei within each ring identically. Down each side of the cylinder, other signal molecules could induce differences in gene expression among different lines but also would treat the nuclei within each line identically. As a result, any nucleus could have a unique pattern of gene activity. In fact, the cylinder is rounded off at each end, and here the mechanism must be modified.

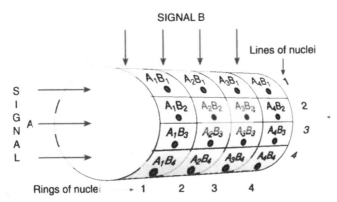

SIGNAL B

Lines of nuclei

SIGNAL A

Rings of nuclei

Both signals form gradients of falling concentration. Nuclei in each ring receive the same concentration of A but different concentrations of B. Those in each line receive the same of B but different of A. Hence each nucleus has a unique environment.

FIGURE 6-1 The principle by which nuclei beneath the surface of a *Drosophila* embryo acquire differences in gene activity. The inducing signals could form gradients of falling concentration (as shown) or act in steps by activating nuclei in each ring or line to emit a further signal.

THE BICOID PROTEIN: A SPECIFIC TRANSCRIPTION FACTOR THAT INITIATES GENE ACTIVATION FROM FRONT TO REAR

Mutations in the 18 genes of *Drosophila* mothers that alter development along the anterior-posterior axis fall into three groups, that is, those that give rise to embryos or larvae with the head and adjacent thorax reduced or absent, with the abdomen reduced or absent, or with the small unsegmented regions at the front and rear (the **acron** and **telson**) altered or absent. It was concluded that molecules laid down in the developing egg within the mother by these 18 genes act in three separate regions of the early embryo. Nüsslein-Volhard, Wieschaus, and other coworkers initially turned their attention to the four maternal genes whose mutation affect head and thorax, and the first fundamental discoveries were made.

The key member of this group of four genes is named *bicoid*.[3] (The thousands of mutations that have been observed in *Drosophila* are each named after the appearance of the mutated embryo, larva, pupa, or fly, and the gene whose change in structure causes the mutation is given the same name.) Mothers that have a mutation in both copies of this gene give rise to embryos with abnormalities that vary from small deletions of the head and thorax to complete absence of these parts, depending on the structure of the protein (if any) formed by the mutated gene (**Fig. 6-2**).

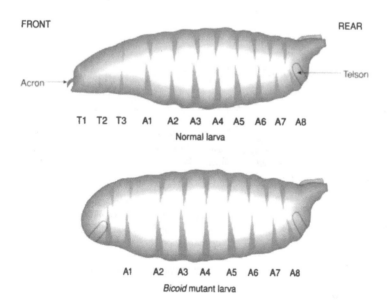

FRONT

REAR

Acron

Telson

T1 T2 T3   A1   A2 A3 A4 A5 A6 A7 A8
Normal larva

A1   A2 A3 A4 A5 A6 A7 A8
Bicoid mutant larva

FIGURE 6-2 Normal and *bicoid* mutant larvae of *Drosophila*. The mutant has segments missing from the head and thorax.

A clue already existed toward explaining the action of the *bicoid* gene: embryos similar to *bicoid* mutants had been reported to develop from normal eggs from which small amounts of cytoplasm had been withdrawn from the front through a glass needle. This suggested that a molecule that initiates development of the head and thorax is concentrated at the front of the egg and might be the molecule altered or missing in *bicoid* mutants. This was confirmed when cytoplasm from the front of a normal egg was injected into the front of one from a *bicoid* mutant mother. Thus, development was wholly or partly restored to normal, depending on the quantity injected. Also, cytoplasm from eggs laid by mothers with an extra copy of the unmutated *bicoid* gene was more effective. Cytoplasm from eggs laid by mothers with one or two copies of the mutant gene was less effective or ineffective, respectively. When cytoplasm from the front of normal eggs was injected in the rear of mutant eggs, a head and thorax formed in these unusual positions. It was concluded that the front of the egg contains molecules that are active in the development of the head and thorax and whose quantity is proportional to the number of unmutated *bicoid* genes in the mother.

Modern techniques enabled the latter molecules to be identified and their method of action understood. The *bicoid* gene, together with adjacent DNA, was isolated, and the nucleotide sequence determined. From this, the sequence of amino acids in the protein formed by the gene was deduced. Messenger RNA was made in the laboratory by transcribing the *bicoid* gene and was used to direct the formation of the bicoid protein. The messenger RNA within the eggs was located and assayed by its hybridization to radioactive *bicoid* DNA. The injection of embryos with fluorescent antibodies against the bicoid protein enabled this protein to be located and assayed in a fluorescence microscope. It was proved that the molecules located at the front of the egg that initiate development of the head and thorax are those of bicoid messenger RNA and that the activity of injected cytoplasm is proportional to its content of these molecules.

The RNA is first detected in **nurse cells**, which cluster around the front of the developing egg within the mother, and it passes from them into the egg. Mutations in two of the three other maternal effect genes that result in defective head and thorax caused bicoid messenger RNA to spread throughout the egg, suggesting that they produce proteins of the microtubules that anchor the RNA. Fertilization provides the stimulus for the translation of the messenger RNA into the bicoid protein, and this protein diffuses through the cytoplasm toward the rear of the embryo (**Fig. 6-3**).

The bicoid protein has a structure characteristic of a transcription factor, and the gene it activates was identified. One of the zygotic mutations that alters the development of the head and thorax is *hunchback*. Embryos that inherit an inactive *hunchback* gene from both parents resemble those from mothers with

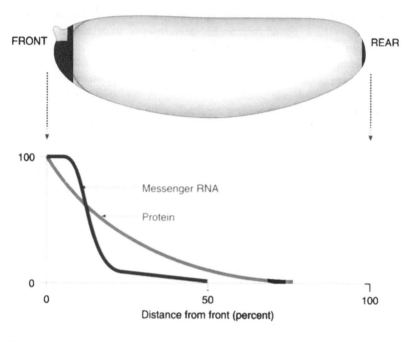

FIGURE 6-3 Distribution of *bicoid* messenger RNA before and after fertilization of *Drosophila* egg and of bicoid protein after fertilization.

inactive *bicoid* genes. This suggests that the bicoid protein activates the *hunchback* gene in certain nuclei that lie beneath the surface of the developing embryo. This theory was supported by the discovery that embryos that lack the bicoid protein also lack the protein formed by the messenger RNA coded by the *hunchback* gene. The theory was confirmed by the following studies of the isolated molecules. The *hunchback* gene plus adjacent DNA was isolated, and its nucleotide sequence was determined. The bicoid protein was found to bind to this DNA and most strongly to three nucleotide sequences near the start of the *hunchback* gene, which were discovered by being protected by the bicoid protein from digestion by pancreatic DNase. It was proved in an elegant way that transcription of an adjacent gene into messenger RNA is activated when the bicoid protein binds to these sequences. The DNA containing these three sequences was separated from the adjacent *bicoid* gene and joined to another gene that forms messenger RNA for a bacterial protein that is more easily assayed (namely, the enzyme chloramphenicol acetyltransferase). The resulting DNA was injected into *Drosophila* embryos, and the bacterial enzyme was later assayed. The quantities were high in normal embryos but undetectable in embryos that lacked bicoid protein, which shows that this protein is needed for activation of the gene. The DNA was also injected into yeast cells. Some of these cells had been engineered to make bicoid protein by injecting the *bicoid* gene. In these cells, but not in the normal yeast cells, the injected DNA formed the bacterial protein. This work leaves no doubt that the bicoid protein is a specific transcription factor that activates the *hunchback* gene in the *Drosophila* embryo by binding to certain adjacent nucleotide sequences.

THE BICOID PROTEIN IS A MORPHOGEN

The bicoid protein diffuses toward the rear of the embryo from its point of formation at the front. The protein is unstable with a half-life of 30 minutes (i.e., at the end of every 30-minute period, half of the molecules diffusing toward the rear of the embryo at the start of the period have been destroyed and replaced by new molecules formed at the front of the embryo). As a result, a concentration gradient in bicoid protein molecules is formed, with maximum concentration at the front of the embryo falling to zero at about one-third of its length from the rear (see Fig. 6-3). As mentioned, the bicoid protein activates the *hunchback* gene to direct the formation of hunchback messenger RNA that is translated into hunchback protein. Assays show that this protein is formed only by nuclei in the front half of the embryo, because only there is the concentration of bicoid protein high enough for it to bind significantly to the nucleotide sequences that activate the *hunchback* gene. The concentration of hunchback protein is constant along the first third of the embryo's length, but falls to zero by half way.

It was now possible to settle a controversial suggestion that had been made repeatedly over many decades,[4] that a concentration gradient of a single compound within an embryo can induce different changes in neighboring nuclei or cells. Molecules that were thought to act in this way were named **morphogens**. Until recently, many biochemists were dubious about this suggestion and preferred to believe that different changes in nuclear or cell composition were induced by different molecules. But the existence of morphogens became increasingly plausible as evidence accumulated that animals can rapidly evolve DNA and proteins with subtle variations in affinity for one another. For instance, a transcription factor might have one amino acid sequence with a high affinity for a regulatory nucleotide sequence in gene A, and another amino acid sequence with low affinity for one in gene B. Hence, the transcription factor could activate genes A and B at a high concentration, and only gene A at a low one.

The discovery that the bicoid protein exists in a concentration gradient made possible the testing of the morphogen hypothesis.[5] The *hunchback* gene has three adjacent nucleotide sequences that bind the bicoid protein. DNA containing some or all of these sequences was isolated and joined to the DNA of a bacterial gene, which makes a protein (β-galactosidase) that is readily detected by a color reaction. This DNA was incorporated into the chromosomes of male and female *Drosophila* flies, and embryos were bred from these flies. In embryos with all three nucleotide sequences linked to the bacterial gene, the distribution of the bacterial protein was similar to that of the hunchback protein in the same embryos, falling to zero concentration at 50% of the length from the front of the embryo. But in embryos with only two or one sequence linked to the gene, the fall in concentration differed from that of the hunchback protein, reaching zero at 40% and 35%, respectively, along the embryo's length.

These experiments provided the first conclusive evidence that an inducing molecule can act as a morphogen, since different concentrations of bicoid protein activated different genes in different nuclei. Thus, in embryos with all three nucleotide sequences preceding the *hunchback* gene but with only one sequence preceding the bacterial gene, the distribution of the two proteins was as follows: nuclei at the front of the embryo formed both proteins, followed toward the rear by nuclei forming only hunchback protein, followed by nuclei forming neither. Although the presence of the bacterial protein is artificial, it was later proved that the bicoid protein also acts as a morphogen when activating *hunchback* and three other zygotic genes. These are named *orthodenticle, button head*, and *empty spiracles* and are activated only by high concentrations of the protein in nuclei at the very front of the embryo. It has also been found that more complex patterns of gene activity can arise when concentration gradients of two transcription factors interact.

Other experiments proved that the group of maternal genes whose mutation results in a defective abdomen (as opposed to head and thorax) have a simple function relating to the formation of this gradient of hunchback protein. The principle gene is *nanos*, whose messenger RNA is anchored at the rear of the embryo; again, the other maternal genes of the group direct the formation of proteins involved in this anchoring. The newly laid egg was found to contain some hunchback messenger RNA evenly distributed within it, and if this RNA were translated into protein at the rear of the embryo, it would prevent the formation of the required concentration gradient in hunchback protein. This translation to the rear is specifically inhibited by the nanos protein. The molecular details have once again been elucidated, but this function was already implied by a simple experiment. Mutant flies whose eggs do not contain the usual hunchback messenger RNA were bred, these producing normal embryos. Hence, the mutant flies should not need to deposit nanos messenger RNA within the egg; in fact, normal embryos are still formed if their *nanos* genes are inactive. Thus, gene activation in embryonic nuclei along the abdomen, as well as the head and thorax, clearly originates from the action of a single protein, namely, bicoid.

CHANGING PATTERNS OF GENE ACTIVITY ALONG THE LENGTH OF THE *DROSOPHILA* EMBRYO

The molecular interactions already described initiate further differences in gene activity along the anterior-posterior axis of the *Drosophila* embryo. The genes that are activated form specific transcription factors that activate more genes to form further specific transcription factors. As a result, the embryo becomes divided along its length into repeating **parasegments**, about four nuclei in length, with certain genes expressed repeatedly in each parasegment. Each segment of the body is formed later from the rear of one parasegment and the front of the adjacent one; hence, segments are out of phase with the earlier parasegments. We now know in considerable detail the molecular interactions that give rise to these changing patterns of gene activity up to the time that the nuclei are enclosed in cell membranes. Some examples are given in this section.

Mutations in at least 27 zygotic genes alter structure along the anterior-posterior axis, and these genes fall into three groups named after the type of structural change. Mutations in 10 **gap** genes, of which *hunchback* is an example, give rise to larvae with up to eight segments missing. Mutations in eight **pair rule** genes give rise to larvae with some particular defect repeated in seven alternate members of the 14 segments. Mutations in nine **segment polarity** genes give rise to larvae with a defect repeated in every segment. Mutations in gap genes were found to be able to prevent the activation of both pair rule and segment polarity genes. Those in pair rule genes leave the activation of gap genes unaffected, but they can prevent the activation of segment polarity genes. However, mutations in segment polarity genes leave both the other groups unaffected. It was concluded that the proteins formed by gap genes activate pair rule genes, and those formed by pair rule genes activate segment polarity genes, thus progressively subdividing the nuclei into smaller groups and building up the similarities and differences between the parasegments.

The activation of many of these genes with the formation of their messenger RNA and protein has been followed over the first few hours after fertilization of the egg. Activation of gap genes begins at about 2 hours, when about 1500 nuclei lie beneath the surface of the embryo, with *hunchback* being the first to be activated in nuclei toward the front of the embryo. By about 4 hours, *hunchback* genes in all nuclei have become inactive, as shown by the disappearance of their messenger RNA. Activation of pair rule genes begins soon after that of the gap genes in almost all nuclei, but by about 3 hours they are active only in seven narrow bands (i.e., rings) of nuclei evenly distributed along the length of the embryo. Segment polarity genes are not activated until after 3 hours when cell membranes are forming around the nuclei. They are active in 14 bands, only one or two nuclei wide. These 14 bands correlate with mutations in these genes, causing defects in each of the 14 segments.

How transcription factors produce this sequence of gene activations has been largely unraveled, primarily by observing how mutations in one gene alter the position along the embryo at which another gene is activated. It is clear that a number of transcription factors are involved in activating or repressing most genes and that many act as morphogens, with different concentrations producing different patterns of activation in different nuclei. This work is illustrated by describing studies of how one of the pair rule genes named *eve* comes to be active in seven repeating bands of nuclei.[6]

It was originally thought that each pair rule gene might be activated by one particular specific transcription factor that recurred seven times along the length of the embryo and bound to a single nucleotide sequence adjacent to the gene. However, mutations in different regions of the DNA adjacent to one of the pair rule genes was found to prevent its activation in different members of the seven bands of nuclei. Hence, different transcription factors appeared to bind to different nucleotide sequences adjacent to the same gene to activate it in each of the seven bands. That the same applies to the pair rule gene *eve* was proved in detail as follows.

The DNA adjacent to the *eve* gene was isolated and linked to the DNA of the gene for the bacterial protein β-galactosidase. This was incorporated into the chromosomes of *Drosophila* flies, and embryos were bred from them. Both the eve protein and the bacterial protein were located in the embryos by their binding labeled antibodies to each protein. Both proteins appeared at the same time in the same nuclei, being finally confined to seven bands, each about two nuclei wide. The DNA adjacent to the *eve* gene was then cleaved into smaller fragments, and these again joined to the bacterial gene and incorporated into the chromosomes of *Drosophila*. Embryos bred from these flies made the bacterial protein in only some of the nuclei that made the eve protein. With one DNA fragment, the bacterial protein appeared only in band 3; with another, only in band 7; with another, only in bands 2 and 7; and with another, only in bands 2, 3, and 7. Clearly, different combinations of transcription factors bind to different nucleotide sequences to activate the *eve* gene in different bands.

Attention was concentrated on the formation of band 2. Since it was found difficult to follow small changes of the eve protein itself in this band, embryos that formed the bacterial protein were studied. When embryos that formed this protein only in bands 2 and 7 had mutations that prevented the formation of either the bicoid or hunchback proteins, the bacterial protein was absent from band 2. This suggested that these transcription factors cooperate to activate the bacterial gene in this band (hence, also the *eve* gene). In embryos that lacked the transcription factor formed by the gap gene *giant*, or that formed by the gap gene *Kruppel*, the bacterial protein in band 2 extended into extra nuclei toward the front or rear, respectively. It was concluded that band 2 is normally only two nuclei wide because the *eve* gene is repressed by the *giant* transcription factor in nuclei to the front and by *Kruppel* transcription factor in nuclei to the rear. In agreement with this, band 2 of eve protein does lie in the embryo between peaks of these giant and Kruppel proteins.

If these conclusions are correct, the DNA fragment that was attached to the bacterial gene in these experiments should contain sequences that bind these transcription factors. These sequences were searched for and found, that is, a stretch of 480 nucleotides contained five sequences that bind bicoid protein and one, three, and three sequences that bind hunchback, giant, and Kruppel proteins, respectively. The activating or repressing action of these four transcription factors on the *eve* gene was confirmed in an elegant way. The sequence of 480 nucleotides was shortened to give rise to a length of DNA with three binding sequences for bicoid protein, two each for giant and Kruppel proteins and one for hunchback protein. This DNA was joined to a bacterial gene for a readily assayed enzyme (chloramphenicol acetyltransferase), and the whole was incorporated into *Drosophila* cells in culture so that they could now make this bacterial protein. Also incorporated into some of the cells were the *bicoid, giant, Kruppel*, or *hunchback* genes. In cells that made bicoid or hunchback proteins, the activity of bacterial enzyme rose by 18 and 4 times, respectively. If the cells made both proteins, the activity rose by 44 times. However, if, in addition, the cells made giant or Kruppel proteins, the activity fell almost to zero (unless these proteins were mutant, without binding ability, when activity was unaffected).

PATTERNS OF GENE ACTIVATION DOWN EACH SIDE OF THE DROSOPHILA EMBRYO

Some mutations in *Drosophila* change the structure of the embryo and larva along the dorsal-ventral axis at right angles to the anterior-posterior axis.[7] In a transverse section of the embryo, although the cells look similar, four regions from dorsal to ventral can be distinguished by their subsequent fate (**Fig. 6-4**). The cells at the top form a membrane that surrounds the embryo, those at the bottom give rise to mesoderm, whereas those at the sides give rise to dorsal and ventral ectoderm. Mutations can alter these fates: for example, the dorsal

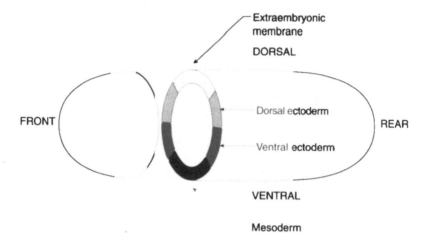

Extraembryonic
membrane

DORSAL

FRONT

Dorsal ectoderm

Ventral ectoderm

REAR

VENTRAL

Mesoderm

FIGURE 6-4 Section through *Drosophila* embryo to show fates within the larva of cells from dorsal to ventral.

ectoderm with its characteristic hairs may extend downward on each side, or the ventral ectoderm may extend upward. These alterations result from changes in specific transcription factors and other proteins that interact to produce this structure and studies, similar to those made on the anterior-posterior axis, have revealed how differences down the dorsal-ventral axis arise.

The series of interactions is complex, but many details have been discovered (**Fig. 6-5**). The series begins during formation of the egg within the mother. It ends in the embryo some hours after fertilization, with the formation of a concentration gradient up each side of the embryo in a transcription factor, within the nuclei, formed by a gene named *dorsal*. This is a morphogen that activates different genes in different nuclei, thus leading to the differences in cell structure down each side of the embryo.

The asymmetric distribution of this dorsal protein, with maximum concentration within nuclei along the lower side of the embryo, derives originally from the nucleus of the unfertilized egg within the mother being asymmetrically positioned (see Fig. 6-5A). A signal protein is formed within the egg nucleus and diffuses through the cytoplasm, but the protein is able to reach only the nearby surface of the egg, which later becomes the upper surface of the embryo. The egg within the mother is surrounded by a layer of **follicle cells**, and the signal protein passes out of the egg to bind to a receptor protein on the surface of follicle cells on this side of the egg. Next, another signal protein from within the egg binds to all follicle cells. Within them, it activates three genes with the formation of three proteins, except in cells that

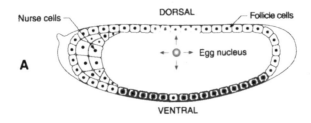

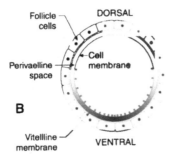

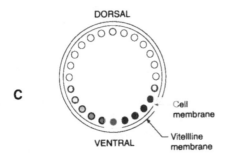

FIGURE 6-5 Steps by which cells of *Drosophila* embryo are directed toward different fates by signals acting down the dorsal-ventral axis. **A:** Longitudinal section of developing egg within mother. Follicle cells surround the developing egg cell and 15 adjacent nurse cells. Egg nucleus releases signal protein (*arrows*), but this reaches only the dorsal follicle cells, where it binds to receptor protein. A second signal (not shown) from within the egg then binds to all follicle cells. Dorsal cells have been inhibited by the previous signal, but ventral cells (*gray*) respond by forming three proteins. **B:** Transverse section of developing egg at later stage. The three proteins give rise to concentration gradient of active signal protein spatzle (*blue*) within the perivitelline space. This, in turn, gives gradient of active *Toll* receptors (*gray dots*) within egg cell membrane. **C:** Transverse section of embryo 3 hours after fertilization. Gradient of active *Toll* receptors in membrane, which now encloses embryo, gives rise to similar gradient of active dorsal protein (*gray*) within embryo nuclei, so directing cells formed later to different fates down the dorsal-ventral axis.

have bound the previous signal protein, which is inhibitory. Hence, these three proteins are confined to follicle cells along what becomes the lower surface of the embryo.

The egg surface is surrounded by two membranes: a tough outer **vitelline membrane** and a fragile cell membrane beneath it. Between the two is a narrow **perivitelline space** filled with fluid. The three proteins pass from the follicle cells through the adjacent vitelline membrane and are confined to the future lower side of the embryo. Here, in the perivitelline fluid, the proteins activate two enzymes that in turn activate another signal protein formed by the gene *spatzle*. This protein diffuses within the perivitelline space to form a gradient of falling concentration toward the future upper surface of the embryo. Lining the inside of the perivitelline space is the egg cell membrane; evenly scattered on its surface are receptors, formed by the gene *Toll*, to which the spatzle protein binds. As a result, a similar gradient of bound *Toll* receptors is formed, their concentration within the membrane gradually falling around each side of the newly laid egg to a minimum on the upper side.

Uniformly distributed within the cytoplasm of the newly laid egg is the protein formed by the *dorsal* gene, but this is inactivated by being bound to another protein. About 3 hours after fertilization of the egg, the *Toll* receptors that are bound to the spatzle protein cause a molecular reaction that removes some of the inactivating protein molecules from the dorsal protein (Fig. 6-5C); they do this most thoroughly where their concentration is highest along the lower side of the embryo. Hence, the concentration gradient in active *Toll* receptors in the membrane is translated into a similar gradient in the active dorsal protein in the cytoplasm adjacent to the membrane. This protein moves into the nuclei. Those nuclei along the lower side of the embryo acquire the highest concentration of dorsal protein, and the concentration per nucleus gradually falls to a minimum at the upper side.

The dorsal protein is a transcription factor, and it acts as a morphogen—activating and repressing different genes at different concentrations—thus giving rise to different types of cell from the lower to the upper side. This has been proved by studies of the dorsal protein's effect on certain genes. Two genes (*twist* and *snail*) are normally active only in nuclei along the lower side of the embryo, suggesting that they are activated by high concentrations of dorsal protein. Two other genes (*zerknullt* and *decapentaplegic*) are active only in nuclei along the upper side, suggesting that they are repressed by dorsal protein. This is supported by mutations that raise the concentration of dorsal protein in all nuclei: they form larvae in which epidermis characteristic of the lower side has moved upward. These mutations cause *twist* and *snail* to be activated in all nuclei and *zerknullt* and *decapentaplegic* to be repressed.

Mutations that lower the concentration of dorsal protein in all nuclei have the opposite effects. These actions of the dorsal protein are supported by studies of isolated genes. The *twist* gene is preceded by nucleotide sequences that bind the dorsal protein, and the *zerknullt* gene is preceded by similar sequences of greater affinity. These genes in various combinations have been introduced into cultured cells. Low concentrations of dorsal protein formed by these cells repress the *zerknullt* gene, but higher concentrations are needed to activate the *twist* gene.

These findings further clarify the way in which morphogens could function. It was clear how a concentration gradient of the bicoid protein within the cytoplasm of the *Drosophila* embryo acts as a morphogen, activating different genes in different nuclei. But it was less clear how a gradient of a signal outside a row of cells might activate different genes in different cells. These findings show how a concentration gradient of a signal protein (spatzle) outside the membrane of the egg cell produces a concentration gradient of activated receptors (Toll) within the membrane, and this produces a concentration gradient in a transcription factor (dorsal) in a series of cell nuclei adjacent to the membrane. Hence, in embryos in which nuclei are separated by cell membranes, a gradient of a signal protein outside a row of cells might well produce a gradient of an activated receptor along the row of cell membranes, and these produce a gradient in a transcription factor from cell to cell, thus leading to the activation of different genes in different cells.

THE FORMATION OF CELLS AND ORGANS OF DIFFERING STRUCTURE IN *DROSOPHILA*

In summary, we now know in great detail the molecular interactions that give each nucleus of the early *Drosophila* embryo its correct pattern of gene activity. The nuclei are evenly distributed over the surface of the embryo but can be envisaged as being composed of groups that foreshadow the later body structure. Thus, the 14 segments along the length of the larval body are foreshadowed by 14 bands of nuclei encircling the embryo from front to rear. The nuclei in each band have certain active genes in common that give rise to the common characteristics of the larval segments. Again, down each side of the larva from above to below there are four regions, each with a distinct type of cell. These regions are foreshadowed by four bands of nuclei along each side of the length of the embryo. The nuclei within each band have certain active genes in common and others that differ.

The cells in which the nuclei become enclosed at first largely differ from one another in their content of transcription factors. These factors direct the formation of proteins in progeny cells that lead to a structure and function appropriate for their position as nerve, muscle, digestive, epidermal, or other cells. Our understanding of the molecular details of these changes is still rudimentary. Clearly important are a group of **homeotic genes**, which are activated by specific transcription factors in the newly formed cells and which themselves code for further specific transcription factors. The importance of homeotic genes was realized many years ago by Edward B. Lewis,[8] who studied their action. For this work, he shared the Nobel Prize in 1995 with Nüsslein-Volhard and Wieschaus.

Homeotic genes in *Drosophila* were discovered by their mutations having dramatic effects on body structure, that is, causing certain large regions of the larva or of the adult fly to develop like other regions. The first mutation was described in 1915 and named *bithorax*. *Drosophila* flies have a pair of rudimentary wings (**halteres**) behind their normal wings, and the *bithorax* mutation causes the front part of each to develop like a normal wing (**Fig. 6-6**) Another mutation was named *Antennapedia*: these mutant flies have pairs of legs protruding from their heads in place of antennae. The positions on the chromosomes of these and other homeotic mutations induced by irradiation were determined. The mutations (hence, the genes) lie in two separate groups on the chromosomes, with those in each group being adjacent to one another. The groups are named the *bithorax* **complex** and the *Antennapedia* **complex**.

The effects of their mutation show that homeotic genes must be involved in initiating the formation of the correct structure of different regions of the body. Although the mutations produce spectacular changes in adult flies, the way the genes act is best deduced from less dramatic effects of their mutation on the segments of larvae. Each segment is covered with a hard cuticle that is secreted by underlying epidermal cells, and its structure is peculiar to each segment. Hence, change from the epidermal cell pattern of one segment to that of another is easily monitored. Lewis[8] studied the effects of various combinations of *bithorax* mutations on segment structure and discovered a strange relationship. When the whole *bithorax* complex was inactivated, a comprehensive alteration occurred. From front to rear,

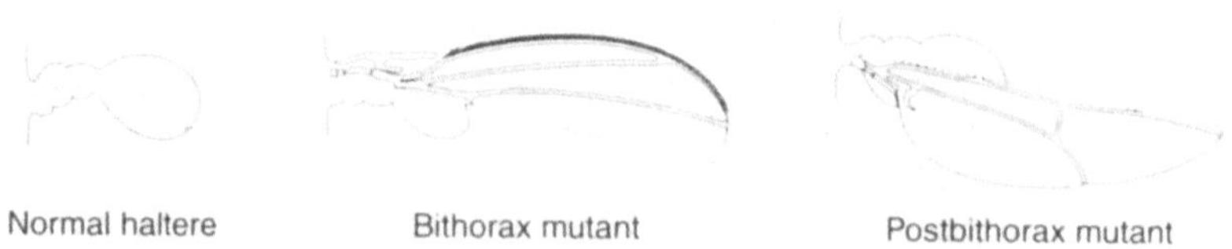

Normal haltere   Bithorax mutant   Postbithorax mutant

FIGURE 6-6 Structure of haltere (vestigial wing) on second thoracic segment of normal adult *Drosophila* and on *bithorax* and *postbithorax* mutant larvae.

the segments were normal through the second thoracic segment, but the third thoracic and all eight abdominal segments had the structure of the second thoracic segment. In larvae with fewer mutations in the *bithorax* complex, the changes in structure were less extreme, and they could be placed in a series of decreasing structural change (**Fig. 6-7**). In the second of the series, the third thoracic segment had returned to normal, and all abdominal segments resembled this segment. In the third of the series, the first abdominal segment had also returned to normal, and the remainder resembled this segment. In the fourth, the second abdominal segment had also returned to normal, and the remainder resembled this segment, and so on, to the last mutant whose segments were all normal except the eighth abdominal, which resembled the seventh.

These findings suggested that in the normal fly the protein formed by one of the *bithorax* genes is present in all cells that will form the third thoracic and subsequent segments. Moreover, this protein alone initiates the formation of the characteristic structure of the third thoracic segment. In addition, an extra protein formed by another gene is also present in cells that will form the fourth and subsequent segments. These two proteins initiate the structure of the fourth segment, and an extra protein is present in cells that will form the fifth and subsequent segments. These three proteins initiate the structure of the fifth segment; and so on, until a combination of six proteins initiates the structure of the eighth segment.

Studies of the *Antennapedia* complex suggested that it determines the structures of the first two thoracic and the head segments in a similar way. To do this, the proteins need to initiate the correct activation or repression of

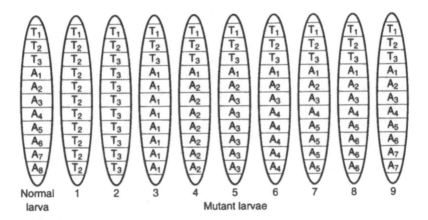

FIGURE 6-7 Arrangement of thoracic (T) and abdominal (A) segments in normal *Drosophila* larva and in series of *bithorax* mutant larvae.

different genes in different cells over the region in which they act. As Lewis[8] concluded,

> The various *bithorax* complex substances are presumed to act indirectly by repressing or activating other sets of genes which then determine the specific structures and functions that characterize a given segment. On such a model the level of development which a segment achieves depends on the particular array of *bithorax* complex substances elaborated in that segment.

The nucleotide sequences of the genes of the *bithorax* and *Antennapedia* complexes were later determined.[9] The structural change was also discovered in the mutant *Antennapedia* gene, which causes adult flies to have pairs of legs protruding from their heads in place of antennae. The promoter that normally binds transcription factors that are present in thoracic segments, thus allowing activation of the gene in these segments, is replaced by another promoter that binds those present in the head. This gave clear evidence that activation of the *Antennapedia* gene initiates the formation of the whole complex structure of the leg. The nucleotide sequences of homeotic genes showed that they all code for protein transcription factors. The *bithorax* complex contains only three genes; hence, the possible combinations of their proteins are insufficient to give the third thoracic and each of the eight abdominal segments of the larva their characteristic structure, as suggested by Lewis. Nevertheless, it seems probable that some modification of Lewis' scheme does apply. The *bithorax* genes are activated by groupings of the proteins of gap and pair rule genes, which differ along the length of the embryo. Moreover, the position of the messenger RNA formed by each *bithorax* gene fits well with a theory that different structures toward the rear of the larva are determined by different combinations of bithorax proteins acting together with some other molecules.

As a result, it first seemed that a simple uniform mechanism, similar to that proposed to act in cell compartments (see Chapter 1), might act throughout *Drosophila*, namely, that a group of cells becomes destined to form a particular region of the body when a particular group of homeotic transcription factors bind in every cell to all genes that will be required to form protein molecules peculiar to the region. Later, each cell receives a personal signal that activates only those genes that are correct for its position within the region. This personal signal might be a particular concentration of a morphogen that forms a gradient across the region.

An important experiment supports the latter theory. Remember that every gene in a cell has an **allele** (a similar gene in an identical position on a homologous chromosome). *Drosophila* embryos were produced, which had a recessive mutation in only one of such a pair of *bithorax* genes and as a consequence did not show any mutant characteristics in their body structure. When embryos are irradiated, chromosome rearrangements occasionally occur during cell division to give rise to a new group (clone) of cells in which a mutant gene is carried on both members of a homologous pair of chromosomes rather than only one. Embryos carrying the *bithorax* mutation were treated in this way, and some formed clones of cells that contained two copies of the mutant gene and so gave rise to the mutant structure. When such clones appeared in the front of the haltere (rudimentary wing) of an adult fly, they, as expected, formed a region with the mutant structure of a normal wing. The important point is this: the precise structure of these cells varied with their position in the haltere and was correct for cells in a similar position in a normal wing. Hence, during development of both the rudimentary and normal wings, cells in similar positions appear to receive the same inducing signals, but they respond differently as a consequence of the prior action of different bithorax proteins. Much research is being done on homeotic genes in *Drosophila* (see Chapter 7), and it has become clear that the way they act is more complex than was at first thought.[9]

There is clear evidence that homeotic genes are also important in other animals. The *bithorax* and *Antennapedia* genes all contain closely related sequences of nucleotides (homeoboxes), which code for **homeodomains** of amino acids that bind to DNA adjacent to other genes. Many animals from *Hydra* to humans have homeobox genes, and they also occur in plants and fungi. Humans and mice have 38 of these genes in four clusters, and, as in *Drosophila*, their order along the chromosome is, for an unknown reason, the same as the order in which their messenger RNAs are formed along the axis of the embryo. Hence, the structures of different regions of the vertebrate body also appear to be determined by different combinations of transcription factors formed by homeotic genes.

REFERENCES

1. Johnston D St, Nüsslein-Volhard C. The origin of pattern and polarity in the *Drosophila* embryo. *Cell* 1992;68:201–219.
2. Kornberg TB, Tabata T. Segmentation of the *Drosophila* embryo. *Curr Opin Gen Dev* 1993;3:585–593.
3. Driever W, Nüsslein-Volhard C. The bicoid protein is a positive regulator of *hunchback* transcription in the early *Drosophila* embryo. *Nature* 1989;337: 138–143.
4. Driever W, Thoma G, Nüsslein-Volhard C. Determination of spatial domains of zygotic gene expression in the *Drosophila* embryo by the affinity of binding sites for the bicoid morphogen. *Nature* 1989;340:363–367.
5. Struhl G, Struhl K, Macdonald PM. The gradient morphogen *bicoid* is a concentration-dependent transcriptional activator. *Cell* 1989;57:1259–1273.

6. Small S, Blair A, Levine M. Regulation of *even-skipped* stripe 2 in the *Drosophila* embryo. *EMBO J* 1992;11:4047–4057.
7. Steward R, Govind S. Dorsal-ventral polarity in the *Drosophila* embryo. *Curr Opin Gen Dev* 1993;3:556–561.
8. Lewis EB. A gene complex controlling segmentation in *Drosophila*. *Nature* 1978;276:565–570.
9. Morata G. Homeotic genes of *Drosophila*. *Curr Opin Gen Dev* 1993;3:606–614.

FURTHER READING

Hoch M, Jackle H. Transcriptional regulation and spatial patterning in *Drosophila*. *Curr Opin Gen Dev* 1993;3:566–573.

Ingham PW. The molecular genetics of embryonic pattern formation in *Drosophila*. *Nature* 1988;335:25–34.

Lawrence PA. *The Making of a Fly*. Oxford: Blackwell Scientific Publications, 1992. (A book devoted to the molecular basis of the development of *Drosophila*.)

CHAPTER SEVEN

THE FRONTIERS OF RESEARCH IN MOLECULAR EMBRYOLOGY

SIGNAL MOLECULES AND THEIR ACTION

In recent years, increasing numbers of scientists have turned to the study of molecular embryology because it contains the major unsolved problems of molecular biology. Many important findings about the molecules that are involved in developmental change are published each year, but, in spite of this, the way in which these molecules interact to convert a fertilized egg into an adult animal is only beginning to be understood. This final chapter is an attempt to summarize the principal topics of current research.

As discussed in Chapter 6, experiments on *Drosophila* showed that differences in gene transcription among nuclei are induced by different specific transcription factors. These factors in the early embryo can pass directly from one nucleus to another. But during the Drosophila embryo's later development and throughout the development of most other animals, interactions among cell nuclei must be mediated by smaller signal molecules that can act across cell membranes. Much research is now devoted to discovering the structure of signal molecules and the **signal transductions** by which they act.

Signal molecules are of two kinds. Some are *soluble* in fat and so can pass through cell membranes; others are *insoluble* and act by binding to protein receptor molecules in the membrane. The fat-soluble group consists mostly of hormones that circulate in the blood and initiate changes in development. Examples are the estrogens that initiate sexual changes in female mammals at puberty. The first important clue to how these molecules act came over 50 years ago, when radioactive tracers became available. Very small quantities of a radioactive estrogen were injected into female rats, and the radioactivity was found to accumulate in the very tissues that the hormone affects, such as the uterus and mammary gland. It was clear that the cells of these tissues contain receptor molecules that bind the hormone. Continuous research since then has revealed the way in which these receptors regulate certain genes after binding the hormone. Like transcription factors, hormone receptors are proteins that have separate regions (**modules**) with different functions. A sequence of amino acids at one end of the molecule can bind the hormone; a sequence in the middle can bind to a regulatory sequence of nucleotides adjacent to a particular gene; and a sequence at the other end of the molecule is able to activate the gene with the cooperation of certain transcription factors. Some receptors lie in the cytoplasm attached to proteins from which they are released when the hormone binds. They then enter the nucleus and bind to the appropriate genes. Other receptors are already bound to the genes but activated only when they also bind the hormone.

Most changes in cell composition during development result from the receipt of inductive signals that are transmitted between closely aligned cells, but few have yet been identified with any certainty. In various animals, many genes whose mutation causes defects in development have been located, and the structures of many of the proteins that they form have been deduced from the nucleotide sequences of the genes. It is clear from these structures that many are growth factors, and many others are their cell surface receptors. Therefore, many inductive signals must be growth factors. These growth factors were originally discovered as compounds, which, in minute quantities, stimulate the growth and development of certain cells.

The first growth factor to be found was **nerve growth factor**. In early experiments, nerve growth factor was injected into newborn mice and unexpectedly was found to stimulate not only nerve cells but also the growth of eyelids. This effect proved to be due to a contaminating **epidermal growth factor**, which was separated and purified. In mammals, this factor induces the differentiation of skin and other epithelia, and its molecular structure and action have been elucidated in detail. Epidermal growth factor is a small protein chain of 53 amino acids and like other proteins cannot penetrate cell membranes. It initiates the transcription of certain genes in epidermal cell nuclei by a series of signal transductions that begin at an **epidermal growth factor receptor**. This receptor is a protein chain of 1186 amino acids that stretches through the cell membrane—one end within the cytoplasm and the other outside the cell and able to bind the growth factor by noncovalent bonds. The binding of the factor causes the receptors to change shape. As a result, they associate in pairs and each receptor catalyzes the addition to its partner of phosphate groups to five amino acids (namely, tyrosines) at the cytoplasmic end of each. This addition changes the cytoplasmic end of the receptor into a shape that will bind a particular protein, which, as a result, becomes able to activate yet another protein, **ras**.

The active ras protein activates another protein by the addition of a phosphate group, and this does the same to yet another protein. Finally, a phosphate group is added to one or more specific transcription factors that become activated to bind to particular genes and to regulate their activity. The many steps in the sequence might appear unnecessary but they have a function: they enable the amplification of the original signal. Such a sequence is called a **cascade**. Thus, although every two molecules of epidermal growth factor that are bound to each pair of receptors on the cell membrane produce only one pair of active receptors, these can activate many molecules of ras protein. These in turn activate many more molecules of the next protein, and so on, until very many molecules of the specific transcription factor are activated. Although only two of these molecules are needed to bind to a homologous pair of genes in each cell, the binding, like all molecular interactions, needs to be sustained by a large reserve of unbound molecules.

The epidermal growth factor receptor is one of a group of closely related **receptor tyrosine kinases** (kinases are enzymes that phosporylate other molecules). Most known growth factors that induce changes in cell composition during development bind to receptors of this group and produce changes in gene activity in a similar way. Each animal uses only a few types of transduction pathways. These are remarkably similar throughout the animal kingdom, having been conserved through millions of years of evolution. Growth factors are also required to regulate cell growth in many adult tissues, and mutations in genes that code for the proteins of the cascades by which they act are a major cause of cancer. The mutated proteins sometimes do not respond to regulatory molecules and so cause uncontrolled cell growth.

RECENT RESEARCH ON *C. ELEGANS*

In recent years, it has become clear that similarities throughout the animal kingdom are greater than was once believed regarding the molecular interactions that cause development. Research workers have concentrated increasingly on three very different animals, *Caenorhabditis elegans*, *Drosophila*, and mice. In these animals, protein molecules that bring about changes in development can be most easily identified as a result of our ability to mutate and isolate genes. Also, many workers are turning their attention to a novel vertebrate, the zebrafish (*Brachydanio rerio*), which has several advantages: a short life cycle, a transparent embryo that enables cell fate to be followed, and easily prepared developmental mutants.

C. elegans was purposely selected in 1974 as a simple animal suited for research on development, and the selection is proving its worth.[1] It has few cells and only about 15,000 genes, and the nucleotide sequence of all its DNA is now known. The development of *C. elegans* is partly mosaic, the different fates of many cells being determined by molecules that are unevenly distributed by asymmetric cell division. However, induction by signal molecules released by one cell and received by another is probably more important. Recent research has identified molecular details of both these processes, largely by searching for mutants with defects in development in the way that has been so successful with *Drosophila*.

Details of inductions by signal molecules have come from studying the development of the vulva, which is the organ through which eggs are laid.[2,3] To identify genes (and thus proteins) involved in its development, research workers induced random mutations in populations of *C. elegans* and searched for individuals with defects in the vulva. Several hundred such individuals have been found with different mutations in nearly 100 genes. The nucleotide sequences of many of the genes have been determined, the structures of the proteins for which they code have been deduced, and many of the proteins have been isolated.

The vulva is composed of 22 cells of various kinds and is formed during the animal's development by the division of three cells within a row of six similar cells with code numbers $P3_p$ through $P8_p$ (**Fig. 7-1**). The central cell of these three ($P6_p$) gives rise to different kinds of progeny cell to the adjacent two. Above this central cell lies an **anchor cell**; if this cell is killed with a laser beam, the vulva does not develop, suggesting that the cell initiates development by releasing one or more signal molecules. If the anchor cell is moved sideways to be centered over any other three of the six cells, these now form the vulva, showing that all six are equivalent and hence are all called **vulval precursor cells**.

A signal molecule released by the anchor cell and many of the proteins that implement its action have been identified. The signal is a protein growth factor, and all vulval precursor cells have a receptor tyrosine kinase that binds it. This is similar to the epidermal growth factor receptor previously described, and the binding initiates a very similar cascade of signal transductions that can result in cell division and development. Inactivation of genes that code for proteins of this cascade prevents formation of the vulva and can cause other defects in development because the same proteins take part in the transduction of signals in other organs. It has been proved that receipt of this signal by the cell beneath the anchor cell ($P6_p$) initiates its division and the formation of the correct progeny cells. But the anchor cell also initiates the different development of the two adjacent precursor cells ($P5_p$ and $P7_p$) to form the complete vulva, and there is conflicting evidence as to how it does this.

One series of experiments suggests that the growth factor is the only signal emitted by the anchor cell and that different concentrations initiate the different development of the precursor cells; that is, it acts as a morphogen. Thus, if all but one of the six vulval precursor cells is killed by a laser beam, the fate of the remaining cell depends on its closeness to the anchor cell. If it is

FIGURE 7-1 Cells involved in development of the vulva of *Caenorhabditis elegans*.

very close, it divides to give rise to the vulval cells normally formed by the central member of the three precursor cells; if it is a little farther away, it gives rise to the vulval cells normally formed by the two outer members of the three; and if still farther away, it does not form vulval cells.

That the growth factor acts as a morphogen is supported by another experiment in which its concentration within the embryos was varied. Like many animals, *C. elegans* has **heat shock** genes, which form proteins that protect the animal from high temperatures. The promoter that activates these genes at high temperatures was joined to the DNA that codes for the growth factor emitted by the anchor cell and introduced into *C. elegans*. As a result, these animals produced throughout their bodies increasing amounts of the growth factor with increasing temperature. In some developing animals of this kind, both the anchor cell and all but one of the six vulval precursor cells were destroyed by a laser beam, and the animals were incubated at different temperatures to produce different levels of the growth factor. At a low temperature, the remaining cell did not form vulval cells; at a higher temperature, it formed vulval cells normally formed by the outer two of the three cells; and at still higher temperatures, it formed vulval cells normally formed by the central cell.

Another experiment confirmed that the vulval precursor cells respond differently to different levels of the signal emitted by the anchor cell. Additional copies of the gene that codes for the growth factor were introduced into *C. elegans* with the result that their anchor cells released more of the growth factor than usual. In these embryos, all vulval precursor cells behaved as if adjacent to the anchor cell.

However, other convincing experiments suggest that on receiving the signal from the anchor cell, the central cell emits a different signal to each of the adjacent cells and that this determines their fate. First, if there is only one signal, development of the two adjacent cells would be expected to be prevented if the cells lack receptors for the growth factor. By gene manipulation, the receptor was deleted from the two adjacent cells but left intact in the central cell. Nevertheless, a normal vulva developed. Although a second signal has not been identified, a gene has been found for an additional membrane receptor (LIN-12 receptor), whose deletion prevents the development of the two adjacent cells. Also, a certain mutation in this gene causes signal transduction to occur when no signal molecule is present, presumably by a change in the structure of the receptor protein. In *C. elegans* with this mutation, all six vulval precursor cells subdivide like the two adjacent cells. It is largely agreed that both sets of experiments have been correctly interpreted and that two different mechanisms operate to initiate development of the two adjacent cells, but why this should be so is unclear. The most popular guess is "to make assurance double sure."

There is also evidence for a third type of signal. The existence of a number of mutant *C. elegans*, in which more than three of the vulval precursor cells divide to form vulval cells, suggests that some molecules that normally prevent this fate are lacking. These mutations fall into two subgroups, and the unusual cell divisions occur only if the embryo has a defective gene in both subgroups. This suggests that there are two molecules that prevent extra cell division. The signals appear to arise in epidermal cells near the precursor cells, but they have not yet been identified. Many other mutations produce subtle changes in development of the vulva, suggesting that the transduction of signal molecules can be modified by molecules involved in other pathways of development. Such complexity in the development of a 22-celled organ in *C. elegans* suggests that far more problems lie ahead in elucidating pathways of development in mammals.

The fate of many cells in *C. elegans* is determined by molecules in the fertilized egg that are distributed to the correct daughter cells by a sequence of asymmetric divisions, and certain cells of the pharynx appear to arise in this way. The two cells formed by division of the fertilized egg (AB and P1; see Fig. 1-4) both give rise to progeny that form pharynx cells. However, if P1 and AB are dissected from the embryo and each is incubated separately in culture, only P1 gives rise to pharynx cells among its progeny. One of the cells formed by division of P1 (i.e., EMS$_l$) and one of the cells formed by division of this cell (i.e., MS$_l$) also gives rise to pharynx cells, as occurs in the intact animal. These facts suggest that formation of the pharynx requires molecules that are present in the fertilized egg and are passed to these daughter cells by asymmetric division to P1 and to the progeny of P1 that form pharynx cells. (The facts also suggest that AB normally gives rise to progeny that form pharynx cells in response to a signal from another cell, and the signal has been proved to come from a descendant of the P1 cell.)

In the hope of discovering molecules passed by asymmetric division to P1 and its progeny,. a study was made of maternal effect mutations that produce animals with too many or no pharynx cells—these cells being easy to identify under a microscope. Deletions of a gene named *skn-1* caused the MS$_t$ cell to form hypodermal rather than pharynx cells. The protein formed by the gene was isolated and found to have the structure of a transcription factor, suggesting that it is involved in activating genes to form an MS$_t$ cell with the correct properties. Deletions of another gene, *mex-1*, caused both cells formed at the first division of the fertilized egg to give rise to progeny with the properties of MS$_t$ cells.

Could it be that the skn-1 protein is normally allotted to the P1 cell after the first division and that the protein formed by the *mex-1* gene is responsible for bringing about this asymmetric distribution? That this is so was elegantly demonstrated.[4] A fluorescent antibody to the skn-1 protein was prepared and embryos were treated with this, thus enabling the position of the protein to be located under a microscope. In normal embryos, the protein was almost entirely located in the nucleus of the P1 cell after the first division of the fertilized egg; however, in embryos in which the *mex-1* gene was deleted, it was evenly distributed between the two nuclei. The asymmetric distribution of the skn-1 protein is thought to be a consequence of the initial asymmetric distribution of some other molecule, possibly the messenger RNA for the protein. The protein is not asymmetrically distributed between the progeny of the P1 cell, suggesting that it is involved in the formation of other cells as well as pharynx cells. The origin of asymmetric cell division by the positioning of the cleavage furrow has already been described. Five *par* genes that affect this positioning have been discovered, and it is hoped that study of the proteins that they form will elucidate how correct asymmetric division is initiated.

Important discoveries are also being made about the function of *Hox* genes in *C. elegans*.[5] The discovery of these genes in *Drosophila*, and their further discovery in a wide range of other animals, has already been described (see Chapter 6). The proteins formed by the genes are transcription factors that cooperate in successive regions from the front to the rear of the animal to initiate the formation of the correct body structure in each region. It was at first thought that the proteins might act in a simple and uniform way; that is, a particular combination of *Hox* genes might produce the same effect in every cell of a particular region of the body, rendering all genes needed for the development of that region susceptible to later activation. Which of these genes were in fact activated in each cell would be decided later by correctly distributed signals, such as concentration gradients of morphogens. Further research has shown that the mechanism of action is less straightforward.

C. elegans contains only one group of four *Hox* genes, each with a structure similar to one of the *Hox* genes in *Drosophila*. In fact, if two of the genes in *C. elegans* are replaced by *Drosophila* genes, they can perform some functions equally well. As usual, the genes mainly influence the development of a particular region from front to rear of the larva and adult, but the way in which they do this is complex and differs from cell to cell and from the way in which they act in *Drosophila*. Sometimes the genes initiate a process characteristic of *C. elegans* development by determining the fate of progeny cells along a certain lineage. Sometimes they initiate certain essential cell migrations. Although the structure of *Hox* genes has remained largely unchanged by natural selection, the ways in which they act appear to have come to differ among animals.

Other current studies on *C. elegans* include the action of signals and receptors in nerve cell formation; the action of genes whose mutation influences structural changes during development, such as invagination of the vulva; and the action of genes whose mutation causes certain changes during development to occur earlier or later than normal. Also identified is a **gerontogene**, whose increased activity increases the life span of the animal by making it more resistant to stresses such as heat and ultraviolet light.[6]

RECENT RESEARCH ON *DROSOPHILA*

Much research is now concentrated on *Drosophila*, inspired by the brilliant discoveries of Nüsslein-Volhard, Wieschaus, and others. The nucleotide sequence of *Drosophila* DNA is now completely known. Particularly successful have been experiments designed to discover signals that convert a uniform sheet of epithelial cells within the eye imaginal disc into the retina of the compound eye.[7,8]

The retina is made up of about 750 groups of exactly 20 cells, each forming a hexagon named an **ommatidium**. At the center of each hexagon, are eight **photoreceptor** cells (nerve cells that respond to light) surrounded by 12 pigment and other cells. The nerve cells can be distinguished from one another and are identified as R1 to R8 (**Fig. 7-2**). So far, attention has concentrated on the formation of these cells.

The eye imaginal disc arises by division of six epithelial cells of the early *Drosophila* embryo to form a sheet, which is one cell thick. Late in the life of the larva, a groove appears at one end of the disc and gradually moves to the other end. As it reaches the end of its passage, it leaves behind clusters of cells that are ommatidia at different stages of formation, and this facilitates the study of their development. The earliest of these clusters each contain a group of three cells one of which, by an unknown induction, becomes the R8 cell, the first nerve cell to be formed. Certain gene deletions make each cell in this group form an R8 cell. This suggests that one cell is randomly directed toward forming the R8 cell, whereupon its genes are activated to form a signal that inhibits the others from following the same fate. This process is called **lateral inhibition** and is used in spacing many other cells in *Drosophila* and in

FIGURE 7-2 The arrangement of photoreceptor cells at the center of each ommatidium of the eye of *Drosophila*.

spacing feathers and hairs in other animals. The inhibition is usually the result of a protein signal, coded by the *Delta* gene, binding to a receptor coded by the *Notch* gene in the inhibited cell.

The next nerve cells to be formed are R2 and R5, which, in the adult eye, are identical in structure and function and differ from R8. The signal that initiates differentiation of the R2 and R5 cells is unknown, but it appears to bind to epidermal growth factor receptors in the cell membrane. These receptors, for reasons not fully understood, are required for development of all the nerve cells of the eye. A specific transcription factor required for the formation of these cells has been identified and is coded by the gene *rough*. The factor is present only in the two precursor cells of R2 and R5, and deletion of its gene prevents their development. R3 and R4 are the next to be formed, followed by R1 and R6. These four cells all differ from one another, but unlike the remaining cells, they all require for their development a transcription factor formed by a gene named (humorously) *sevenup*. Another transcription factor (Bar) is confined to R1 and R6 and is also essential for their development.

Most is known about the formation of the R7 cell. In 1976, a mutant *Drosophila* was found in which this cell developed abortively; it was named *sevenless*. The gene was isolated and shown to code for a receptor tyrosine kinase. Deletion of another gene, active only in the adjacent R8 cell, had an identical effect; it was named *bride of sevenless* (*boss*). This gene was also isolated and its protein proved to be the signal that activates the receptor in the R7 cell, thus inducing its development. This signal is unusual in both structure and action. It has the structure of a somewhat unusual membrane receptor and lies on the surface of the R8 cell. Its method of action was studied on genetically manipulated cells in culture. When cells with an active *sevenless* gene were mixed with others with an active *bride of sevenless* gene, they clumped together owing to mutual binding of the proteins formed by the two genes. In so doing, the sevenless protein became phosphorylated, which is the normal way in which such receptors are activated. It is concluded that, when the R7 and R8 cells make contact, the bride of sevenless protein moves into the R7 cell and binds to its receptor, thus initiating signal transductions that lead to the activation of transcription factors. The R1–R6 cells also contain the sevenless protein and are also in contact with R8 cell, but they do not develop as R7 cells. The reason is that they do not engulf the bride of sevenless protein, apparently as a consequence of having already followed other paths of development.

Other recent experiments in *Drosophila* have been designed to discover the inductions that determine the different structures of the fore and hind parts of the adult wing. As already described, the wings arise from wing discs of the embryo that are composed of fore and hind **compartments** of cells of different lineage, with the gene *engrailed* active in cells of the rear compartment only. The formation of differences in wing structure along the anterior-posterior axis depends on the secretion of a protein growth factor (in response to signals from the rear cells) by a narrow band of cells along the boundary between the front and rear compartments. This growth factor (product of the gene *decapentaplegic*) diffuses toward both front and rear of the wing disc to form two concentration gradients. It is thought that different concentrations induce cells with the correct differences in composition and structure and that the different responses of cells to the rear arise from their each having an active *engrailed* gene.[9,10]

If the latter interpretation is correct, then the mechanism is similar to that proposed when compartments were first discovered. The *engrailed* gene contains a homeobox, and it was proposed that the different homeotic genes act in a similar way during the formation of differences in structure along the anterior-posterior axis of the body. Much research has recently been done on homeotic genes in *Drosophila*, but the many facts that have accumulated have not revealed such a simple common mechanism of action.[11,12] On the contrary, their action appears complex and not identical with that in other organisms. Homeotic genes are first activated by transcription factors formed by *gap, pair rule*, and other genes, and this activation is partly maintained by autocatalysis by the transcription factors formed by the homeotic genes themselves. Homeotic proteins are present throughout development in each region that they regulate, but they exist at different concentrations in different cells. They bind to a sequence of six nucleotides in DNA, but changes in several of these nucleotides cause only small changes in affinity. They are thus much less particular than most transcription factors about the precise DNA sequence to which they bind, and they appear to cooperate with other molecules to bind in a highly adaptable way. As a result, each factor binds to many genes, and more than one factor will bind to the same gene. Hence, it is often difficult to conclude which genes are directly regulated by homeotic proteins, but they appear to total over 1000. Such genes include those that code for other transcription factors, signal molecules, membrane receptors, and also structural proteins of cells, indicating that homeotic proteins have different functions in different cells. Clearly, much work remains to be done to clarify the obvious importance of homeotic genes.

Progress has recently been made in discovering the components of pole plasm that direct the formation of germ cells.[13,14] The influence of cell adhesion molecules on the formation of body structure in *Drosophila* is also being studied. Mutations have been induced in integrin genes, and the first *Drosophila* cadherin molecule has recently been identified.

RECENT RESEARCH ON *XENOPUS*

I have already described experiments on amphibia, which reveal inductions that initiate the formation of body structure, namely, the induction in blastulae of various types of mesoderm cell by vegetal epidermal cells and the induction in gastrulae of nerve cells by underlying mesoderm cells. Many experiments are now being done to discover the mechanism of mesoderm induction in *Xenopus*, and it is becoming clear that the inductions are more intricate than proposed by the simple three-signal model.[15-17] The inducing signals appear to be protein growth factors whose action is transmitted by their binding to specific receptors on cell membranes. Thus, blastula cells have been cultured in tiny wells on plastic plates, and different growth factors have been added, several of which induce the formation of molecules characteristic of particular mesoderm cells. Also, messenger RNAs that will direct the formation of certain growth factors have been injected into early embryos in which mesoderm formation has been inhibited by ultraviolet radiation, and these RNAs restore the formation of mesoderm cells. This technique is often simpler than injecting the growth factor itself, because the messenger RNA is more easily obtained pure.

Ideally, if an active factor is to be accepted as the true inducing molecule, three criteria should be satisfied: (1) it should cause the correct induction in one of these tests; (2) it should be found in the embryo in the correct position; and (3) when its action is inhibited, induction should stop. Few growth factors fulfill all these criteria, but several come near it. The first strong candidate was a growth factor named **noggin**, which was identified by injecting pure messenger RNAs isolated from *Xenopus* gastrulae into irradiated embryos. Noggin induces epidermal cells of blastulae to form mesoderm cells, and its messenger RNA is confined to the correct region of the blastula. However, a method of specifically inhibiting noggin action has not yet been devised. Since then, some 20 other growth factors have been found to restore mesoderm formation in irradiated embryos, the most active being **activin**.

One way of testing the importance to the embryo of a growth factor is to inject the messenger RNA for a **dominant negative receptor**. The protein formed is a mutant version of the normal receptor, and it inhibits the action of the growth factor. Formation of such a receptor for activin delays mesoderm induction. Activin is found to be correctly placed in *Xenopus* embryos and so fulfils all three requirements for a true signal. Interesting experiments suggest that activin may function as a morphogen in mesoderm induction, a gradient of falling concentration inducing different types of cell from the same type of blastula cell. Animal hemispheres from *Xenopus* blastulae were dissociated into single cells and incubated with minute amounts of activin. Different types of mesoderm cell could be induced with each increase in activin concentration of 1.5 times. It might be argued that this is because the hemispheres are composed of cells of different types that respond differently to activin. However, this was disproved by further experiments in which small beads loaded with activin were implanted into intact animal hemispheres.[18] The resulting expression of genes characteristic of mesoderm cells was detected in microscopic sections by hybridization of labeled probes to the messenger RNA of these genes. Different concentrations of activin induced the expression of different genes, and all adjacent cells behaved similarly. Progress is being made toward discovering the molecular pathways by which activation of signal receptors induces the transformation of cells into mesoderm and the transcription factors involved. The terminology is complex, as shown by the following heading to a paragraph in a recent review, "What is the relationship between the Xwnt/beta-catenin/XTcf3 dorsal signaling pathway and the Smad2/FAST pathway?"

Many laboratories are now studying the molecular basis of neural induction in *Xenopus*.[19–21] This animal appears to be better for this purpose than *Triturus*, since its ectodermal tissue is less liable to respond to the improbable inducers that so confounded early experimenters. The gradient model of Toivonen and Saxen (see Chapter 4) has been a popular guide, but reality is more complex. Detection of newly induced nerve tissue has been made precise by assaying molecular markers that occur only in this tissue; a useful marker is messenger RNA for the neural cell adhesion molecule (N-CAM). Excised ectoderm from gastrulae has been incubated with many growth factors in the hope of inducing nerve tissue. The protein noggin described earlier and some other growth factors have given clear results: when they are incubated with ectoderm from a late blastula or gastrula, messenger RNA for N-CAM is formed. Factors of another type, bone morphogenetic proteins, are also involved, and their properties have suggested a modified mechanism for neural induction: molecules needed for the transition of ectoderm cells to neural cells are induced at an earlier stage but are inhibited by the binding of bone morphogenetic proteins to cell receptors. Growth factors released by underlying mesoderm, such as noggin, initiate the transition by removing this inhibition. Previously, one difficulty of working with *Xenopus* was that modification of chosen genes was impossible, but this is now achieved by modifying the DNA of sperm before their introduction into unfertilized eggs. In this way, frogs that form no fibroblast growth factor have been produced. Their neural induction is unaffected, suggesting that this particular factor is not involved, although other growth factors that are closely related appear to be.

As described in Chapter 3, Spemann believed that neural tissue arises partly from inductions that move forward cell by cell along the ectoderm from the dorsal lip, an idea dismissed by his contemporaries. Work over the last few years suggests that Spemann was correct. The principal evidence against this theory was that exogastrulae make no nerve tissue, but recently *Xenopus* exogastrulae were shown to form neural cell adhesion molecules in ectoderm at its junction with mesoderm. Studies were then made on pieces of ectoderm plus adjacent dorsal lip taken from embryos before gastrulation and maintained in culture. The explants were kept flat to prevent mesoderm of the dorsal lip from sending signals to ectoderm from below. The ectoderm formed not only neural cell adhesion molecule but also some neural structures. Clearly, in addition to reaching ectoderm from underlying mesoderm, inducing signals do in fact move forward along the ectoderm from the dorsal lip. This mechanism appears to become important only after the initial stages of neural induction.

Late in neural induction, a complex series of changes occurs within the nervous system, which ends in the formation of junctions between individual nerve cells. Progress is being made in discovering their origin. Many of these structural changes are prevented in the developing neural tube by removing underlying tissue (notochord). This tissue appears to release two signals that induce the floor plate of the neural tube. The floor plate in turn releases signals that interact with others released by the tube to produce additional changes in its structure. A number of protein growth factors appear to be involved, including the Sonic hedgehog protein released by the notochord. Clearly, inductions that induce the nervous system are complex, and much work is needed to unravel them.

Experiments are also being done on *Xenopus* to study the functions of cell adhesion molecules in development. For example, in a recent experiment the gene for C-cadherin, which is the principal cell adhesion molecule in the blastula, was isolated and mutated by removing part of the DNA. RNA transcripts of the mutated gene were prepared and injected into those cells of blastulae that form the involuting dorsal lip of the blastopore at gastrulation. The transcripts became translated into a mutant protein, which acted as a dominant negative antagonist of C-cadherin. The effects were remarkably specific: the embryos were unable to complete the cell movements of gastrulation. If an excess of the normal messenger RNA for C-cadherin was injected with the mutant RNA, gastrulation returned to normal.[22]

RECENT RESEARCH ON MICE

Our ability to induce mutations in *C. elegans* and *Drosophila* and to select individuals with defects in development has given much information about the molecular basis of development. However, mice are not suited to similar procedures. For some years, it has been possible to introduce foreign genes into unspecified positions in the chromosomes of mice by injecting the DNA into the nuclei of fertilized eggs. But the study of mice has been revolutionized by discovering almost the complete nucleotide sequence of their DNA and by the technique of gene targeting or gene knockout, by which a chosen gene can be modified or inactivated (see Chapter 5). Although gene targeting is a new technique, it has already provided important information. One surprising discovery is that deletion of some active genes has little or no obvious effect (although it is difficult to believe that they are surplus to requirements). Gene targeting is also providing facts about the action of homeotic genes and about the molecular interactions that induce the development of kidney, brain, bone, and epidermis.[23]

Mice have 39 homeotic genes arranged in four clusters. These genes are very similar in structure to the homeotic genes of *Drosophila*, and some of the mouse genes even replace the *Drosophila* genes satisfactorily. As in *Drosophila* (see Chapter 6), the body regions in which the different genes are transcribed into RNA suggest that they act in different groupings to initiate the formation of different regions along the anterior–posterior axis. Only two genes are active at the front of the mouse embryo, whereas increasing numbers are active toward the rear, where finally all are active. The order of gene expression from front to rear once again corresponds to their order along the chromosome. The boundary of a region in which an additional gene is expressed often corresponds to a structural boundary within the animal, such as the boundary between cervical and thoracic vertebrae. Mice have now been produced with mutations in each of the homeotic genes, and, as expected, deletion of a particular gene often produces defects in the development of a particular limited region of the body along the anterior-posterior axis. As in *Drosophila*, the defective region often resembles the immediately preceding region. For example, deletion of one homeotic gene causes the first lumbar vertebra to form a rib like the preceding thoracic vertebra. But little is known about how the transcription factors that are formed by these homeotic genes act.

The genes for cell adhesion molecules, which are now thought to be one of the main agents that determine body shape, would appear to be likely targets for the transcription factors formed by homeotic genes. This interaction is now being studied in mice. It has been shown that homeotic proteins do bind to nucleotide sequences adjacent to genes for cell adhesion molecules, and experiments on cultured cells show that homeotic proteins can regulate the formation of adhesion molecules. The effects of the knockout of genes for cell adhesion molecules on the development of mice is also being studied. For example, embryos with the N-cadherin gene deleted die with malformations of the heart.

RECENT RESEARCH ON LIMB DEVELOPMENT

Remarkable progress is being made in elucidating the mechanism of limb development.[24-26] Experiments have continued on chick embryos and have been extended to mice, in which gene knockout has given important information. Techniques have also been developed to introduce genes into chicks by attaching them to an infecting virus.

Harrison showed years ago that extra forelegs or hind legs can be induced at any point along the side of a newt embryo by implanting fragments of developing ear or nose beneath the mesoderm (see Chapter 4). It has now been shown that normal legs and wings can be similarly induced in chick embryos by implanting beads soaked in a solution of chick fibroblast growth factor 8, suggesting that this, or some similar growth factor, initiates the formation of limb buds. The importance of the apical ectodermal ridge at the tip of the limb bud has already been described, and fibroblast growth factors also appear to be involved in its action. Different factors are located in different cells of the ridge, and if the ridge is removed, the addition of growth factors maintains outgrowth of the limb.

Limb development is determined by three largely independent sets of inductions along three axes: anterior-posterior, dorsal-ventral, and proximal-distal. Tissue grafting showed that mesoderm cells at the rear of the wing bud of a chick embryo are the source of signals that determine structure along the anterior-posterior axis of the wing (see Chapter 4). Attempts were made to find a pure compound that would mimic the action of these mesoderm cells. Pieces of paper that had been dipped into solutions of various compounds were inserted into slits at the front of wing buds, and one compound—retinoic acid—was found to give spectacular results. Retinoic acid induced the formation of a second wing just as if mesoderm cells from the rear of a limb bud had been implanted in the same position. Retinoic acid is formed in animals from vitamin A in food. Being soluble in fat, it can pass through cell membranes into the cytoplasm, where, after binding to a specific receptor, it can activate genes that code for transcription factors. It is present in limb buds and concentrated in the rear mesoderm cells. Retinoic acid receptors are also present on membranes of mesoderm cells in limb buds. Hence, it was first thought that retinoic acid is secreted by mesoderm cells at the rear of a limb bud and acts as a morphogen, with different wing structures from posterior to anterior induced by different concentrations of retinoic acid.

However, it has been proved that implanted retinoic acid acts indirectly by inducing neighboring cells to release a further signal that is normally confined to the rear mesoderm cells. This is a protein growth factor coded by the gene *Sonic hedgehog*. When beads soaked in solutions of the growth factor are implanted at the front of a limb bud, they produce the same transformations as rear mesoderm cells. If mesoderm cells that contain the protein are excised, development along the anterior-posterior axis is defective, as it is in mice in which the *Sonic hedgehog* gene has been inactivated. It is improbable that the protein acts as a morphogen, because it does not diffuse from the cells that produce it and hence must activate further signals. The initial activation of the gene appears to result from a pattern of *Hox* gene activation, which is in turn regulated by retinoic acid.

Progress has also been made toward discovering the signals that determine limb structure along the dorsal-ventral axis. Differences in structure along this axis appear to be initiated by ectoderm cells. Messenger RNAs formed by many genes involved in cell signaling have been located in the ectodermal cells of the limb bud. They are unevenly distributed, and different genes are expressed on the dorsal and ventral sides of the apical ridge. By gene manipulation, these genes have been overexpressed in chicks and deleted in mice. Overexpression of a gene confined to one side of the apical ridge can cause the structure typical of that side to extend to the other side, whereas its deletion can cause the reverse. Hence, growth factors and transcription factors formed by these genes appear to be involved in determining structure along the dorsal-ventral axis. Little new evidence exists as to how structural differences along the proximal-distal axis are induced except that fibroblast growth factors are important, as shown by their ability to restore limb growth after the apical ridge is removed.

AFTERTHOUGHTS

It is possible for research in some branches of science to come to an end through virtually all facts being discovered. However, research in animal development is clearly in its infancy, and important molecular strategies await discovery. For example, how are the effects of mutation in homeotic genes to be explained? What is the explanation of the coordinated behavior of cells in a developmental field? Why are epidermal cells of insects stimulated to divide when placed beside apparently identical cells from a different region (see Chapter 4)? What is the significance of compartments? Also, there is growing interest in the relation between development and evolution.[27,28] For example, mutations that influence development can produce dramatic changes in body structure that might be favored by natural selection. Clearly, no research worker entering animal development need fear a shortage of problems.[29]

REFERENCES

1. Wood WB, Edgar LG. Patterning in the *C. elegans* embryo. *Trends Genet* 1994;10:49–54.
2. Kenyon C. A perfect vulva every time: gradients and signaling cascades in *C. elegans*. *Cell* 1995;82:171–174.
3. Sundaram M, Han M. Control and integration of cell signaling pathways during *C. elegans* vulval development. *BioEssays* 1996;18:473–479.
4. Bowerman B, Draper BW, Mello CC, Priess JR. The maternal gene *skn-1* encodes a protein that is distributed unequally in early *C. elegans* embryos. *Cell* 1993;74:443–452.
5. Kenyon CJ, Austin J, et al. The dance of the *Hox* genes: patterning the anteroposterior body axis of *Caenorhabditis elegans*. *Cold Spring Harb Symp Quant Biol* 1997;LXII:293–305.
6. Murakami S, Johnson TE. Life extension and stress resistance in *Caenorhabditis elegans* modulated by the *tkr-1* gene. *Curr Biol* 1998;8:1091–1094.
7. Cagan R. Cell fate specification in the developing *Drosophila* retina. *Development* 1993;(Suppl):19–28.
8. Dominguez M, Wasserman J.D, Freeman M. Multiple functions of the EGF receptor in *Drosophila* eye development. *Curr Biol* 1998;8:1039–1048.
9. Nellen D, Burke R, Struhl G, Basler K. Direct and long-range action of a *dpp* morphogen gradient. *Cell* 1996;85:357–368.
10. Lecuit T, Brook WJ, et al. Two distinct mechanisms for long-range patterning by Decapentaplegic in the *Drosophila* wing. *Nature* 1996;381:387–393.
11. Biggin MD, McGinnis W. Regulation of segmentation and segmental identity by *Drosophila* homeoproteins: the role of DNA binding in functional activity and specificity. *Development* 1997;124:4425–4433.
12. Akam M. *Hox* genes: From master genes to micromanagers. *Curr Biol* 1998;8:R676–R678.
13. Williamson A, Lehmann R. Germ cell development in *Drosophila*. *Ann Rev Cell Dev Biol* 1996;12:365–391.
14. Wilson JE, Macdonald PM. Formation of germ cells in *Drosophila*. *Curr Opin Genet Dev* 1993;3:562–565.
15. Harland R, Gerhart J. Formation and function of Spemann's organiser. *Ann Rev Cell Dev Biol* 1997;13:611–667.
16. Heasman J. Patterning the *Xenopus* blastula. *Development* 1997;124:4179–4191.

17. De Robertis EM, Kim S, et al. Patterning by genes expressed in Spemann's organiser. *Cold Spring Harb Symp Quant Biol* 1997;LXII:169–175.
18. Gurdon JB, Ryan K, et al. Cell response to different concentrations of a morphogen: activin effects on *Xenopus* animal caps. *Cold Spring Harb Symp Quant Biol* 1997;LXII:151–158.
19. Harland RM. Neural induction in *Xenopus*. *Curr Opin Gen Dev* 1994;4:543–549.
20. Doniach T. Basic FGF as an inducer of anteroposterior neural pattern. *Cell* 1995;83:1067–1070.
21. Grunz H. Neural induction in amphibians. *Curr Top Dev Biol* 1997;35:191–228.
22. Lee C-H, Gumbiner BM. Disruption of gastrulation movements in *Xenopus* by a dominant-negative mutant for C-cadherin. *Dev Biol* 1995;171:363–373.
23. Capecchi MR. *Hox* genes and mammalian development. *Cold Spring Harb Symp Quant Biol* 1997;LXII:273–281.
24. Schwabe JWR, Rodriguez-Esteban C, et al. Outgrowth and patterning of the vertebrate limb. *Cold Spring Harb Symp Quant Biol* 1997;LXII:431–435.
25. Johnson RL, Tabin CJ. Molecular models for vertebrate limb development. *Cell* 1997;90:979–990.
26. Cohn MJ, Tickle C. Limbs: a model for pattern formation within the vertebrate body plan. *Trends Genet* 1996;12:253–257.
27. Raff RA, Kaufman TC. *Embryos, Genes and Evolution*. New York: Macmillan, 1983.
28. Gerhart J, Kirschner M. *Cells, Embryos and Evolution*. Oxford: Blackwell Science, 1997.
29. Fraser SE, Harland RM. The molecular metamorphosis of experimental embryology. *Cell* 2000;100:41–45.

FURTHER READING

Pattern Formation During Development. *Cold Spring Harb Symp Quant Biol* LXII, Cold Spring Harbor Laboratory Press, 1997. (A good summary of recent problems and ideas.)

GLOSSARY

allele   originally meant the alternative forms of a mendelian unit character, such as red versus white flowers; now also applied to the alternative structures of the gene that directs the formation of the character.

animal development   the gradual increase in complexity of an embryo, then of a young animal, to form an adult; the term is often used instead of "embryology," since it also embraces the study of juvenile animals including larvae.

animal hemisphere   the half of the egg cell with the animal pole at its center.

animal pole   the region of an egg cell beneath which lies the cell nucleus. Nutritive yolk granules are often concentrated beneath the opposite vegetal pole.

amino acids   20 small molecules of similar structure that living cells join together by covalent bonds into long chains to form proteins.

anterior-posterior axis   the line from front to rear of an animal.

apoptosis   programmed cell death.

archenteron   new cavity formed by the invagination of the blastula during gastrulation.

atoms   particles that compose each of the 100 or so elements (e.g., oxygen, carbon, silver), the atoms within any one element being identical. When identical or different atoms join together by chemical bonds, they form molecules.

axis   see *dorsal-ventral axis* and *anterior-posterior axis*.

base pairing   the noncovalent bonding between the complementary bases (guanine and cytosine, and adenine and thymine or uracil), which stabilizes the double-stranded conformation of nucleic acids.

blastocoel   the cavity within a blastula.

blastopore   the indentation that forms in a blastula when cells invaginate to form the gastrula.

blastula   a hollow sphere of cells formed during the early cell divisions of an embryo; in many animals, this basic structure, which is clear in sea urchins and frogs, is greatly modified.

catalyst   an element or compound that accelerates a reaction between other molecules, but itself emerges unchanged after the reaction is complete. Enzymes are the principal catalysts of living cells.

cell cycle   the period from the formation of a new cell by division of its parent cell until the division of this new cell into two.

cell fate   the "fate" of a particular embryonic cell is to give rise to a particular clone of progeny cells.

cell line   abnormal cells, which, when separated from animal tissues divide indefinitely in culture, such as cells from cancerous tissues.

cell lineage   the line of a cell's ancestral cells going back to the fertilized egg.

cells   units of which most animals (and plants and bacteria) are composed; see also individual names of cells.

centrosome   a structure located at one side of the cell nucleus from which microtubules radiate. Before cell division, the centrosome splits into two centrosomes, which separate to opposite poles of the cell, and the cleavage furrow forms halfway between them.

chromatid   see *mitosis*.

chromatin   the material of which chromosomes are composed.

chromosomes   threads within cell nuclei that become visible in a light microscope when the nuclear membrane disintegrates before cell division;

usually composed of one long DNA double helix, which is coiled into a superhelix around protein particles called nucleosomes, with many other proteins adhering by weak chemical bonds. Chromosomes carry genes, each being a particular portion of the length of this double helix.

cleavage divisions   a sequence of rapid, often synchronous, cell divisions by which a fertilized egg is subdivided (without growth) to form the blastula.

cleavage furrow   constriction that forms around the surface of a dividing cell and finally separates it into two cells.

clone   a population of cells formed from a single cell by cell division.

compound   a molecule composed of more than one kind of atom.

conformations   the three-dimensional structures that a molecule can assume as a result of the rotation of one group of atoms relative to another about single covalent bonds. DNA occurs in two distinct conformations: double helixes, which are stabilized by noncovalent bonds between an adjacent pair of molecules (or "strands"), and single molecules in unstabilized "random coils." The properties of a protein depend on the particular stable conformation it assumes as a result of noncovalent attractions and repulsions between the amino acids of its covalent or primary structure.

contractile ring   a ring of actin filaments beneath the surface of a dividing cell, which contract to form the cleavage furrow.

cortex   a dense network of actin filaments and other proteins that lie beneath the cell membrane; the cortex gives strength to the cell and directs changes in its shape and movement.

covalent bond   the commonest and strongest of the bonds that link atoms together to form molecules; three other weaker kinds of bond are noncovalent bonds.

cytoplasm   the viscous fluid within cells that surrounds the nucleus and other cell structures and contains protein and other molecules in solution; originally called protoplasm.

cytoskeleton   a network of contractile filaments within cells that control their shape and movement.

denaturation   the loss of biological activity of a molecule, resulting from a permanent change in its conformation.

determinants   molecules that can be asymmetrically distributed at cell division and that determine the fate of cells that receive them.

differentiation   process in which differences in molecular composition—hence in appearance and behavior—arise between cells and tissues during development.

DNA   deoxyribonucleic acid; DNA molecules are formed by covalent bonding into long chains of four small nucleotides named deoxyadenylic acid, deoxythymidylic acid, deoxyguanylic acid, and deoxycytidylic acid (usually abbreviated to A, T, G, and C). The differences between DNA molecules depend on the order in which the four different nucleotides are arranged along the chain and the total number in the chain. In living cells, the chains (or "strands") occur in complementary pairs of equal length held together by noncovalent bonds between successive pairs of nucleotides. The whole structure is coiled into a helix.

dominant and **recessive**   terms introduced by Mendel to describe the alternative forms of a unit character. The terms are now also applied to the alternative molecular structures of the pair of genes that control the character. Recessive genes usually code for no protein, or one of reduced activity.

dorsal-ventral axis   the line between the upper and lower sides of an animal perpendicular to the anterior-posterior axis; derived from the Latin *dorsum* for back and *ventrum* for belly.

ectoderm, endoderm, and **mesoderm**   the three cell layers that are formed at gastrulation.

egg   the cell contributed by a female that unites with the sperm cell contributed by a male to form the fertilized egg cell (zygote). Egg and sperm cells have only one copy of each chromosome and hence one copy of each gene.

element   every substance is composed of the atoms of elements of which there are around 100 (e.g., carbon, hydrogen, phosphorus, iron).

embryo   first formed when an egg cell is fertilized by a sperm. The developing animal ceases to be called an embryo after birth or hatching from an egg.

embryology   the study of all aspects of the growth and development of embryos.

embryonic induction   change in molecular composition of a layer of embryonic cells in response to a signal received from another closely aligned cell layer.

endoderm   see *ectoderm*.

endonuclease   an enzyme that severs a nucleic acid molecule into two polynucleotides.

enhancer   a nucleotide sequence adjacent to a gene that binds specific transcription factors that regulate the activity of the gene.

enzyme   a protein that catalyzes a particular molecular reaction within a cell. Changes in amino acid sequence of the enzyme, as a result of mutations in its gene, change its conformation and hence its activity.

epidermis   see *epithelium*.

epithelium   a layer of adhering epithelial cells, one cell thick; epithelia often line the outer (*syn.*, epidermis) and inner surfaces of an animal's body.

exonuclease   an enzyme that removes a single nucleotide from the beginning or end of a nucleic acid molecule.

extracellular matrix   extracellular molecules, mainly proteins and polysaccharides, which form an increasing part of an animal's body as development proceeds. The matrix gives strength to the body and provides a surface to which cells adhere.

fibroblasts   members of the family of connective tissue cells that secrete collagen and other molecules of the extracellular matrix. They migrate through the body and lodge in wounds that they help to repair; they are the easiest vertebrate cells to grow in culture.

gastrula   the structure formed from the blastula by gastrulation.

gastrulation   the indentation or invagination of the blastula to give rise to three layers of cells (ectoderm, endoderm and mesoderm) from which different body tissues develop.

gene   a discrete length of the long DNA double helix within a chromosome. Genes perform two basic functions in each cell: (1) they direct the formation of exact copies of themselves, thus ensuring that the new cells formed by division inherit the correct genes; and (2) each gene directs the formation of a particular RNA molecule, with a nucleotide sequence that is complementary to that of one of its two strands. Most of the RNA molecules are (or are converted to) messenger RNAs, each of which directs the formation of a particular protein. These proteins determine the molecular reactions of the cell and, as a result, genes control the formation or entry into the cell of every cell molecule.

germ cell   egg or sperm cell.

germ plasm   cytoplasm that induces the formation of germ cells.

growth factor   proteins that act in minute concentration to induce changes in cell composition.

halteres   vestigial wings of *Drosophila*.

helix   a corkscrew shape.

heterochromatin   condensed, dark-staining regions within chromosomes in which inactive genes are often located.

heterozygous   see *homozygous*.

histoblast nest   groups of cells within the larva of *Drosophila*, which contribute to the structure of the adult fly; see also *imaginal discs*.

histones   five proteins that maintain chromosome structure; see *nucleosome*.

homeoboxes   a group of related nucleotide sequences, 180 base pairs long, which were first identified in homeotic genes of *Drosophila*. They code for sequences of 60 amino acids known as a homeodomains in homeotic protein transcription factors. Homeodomains all have similar conformations and are responsible for binding the transcription factors to the genes they regulate.

homeodomains   see *homeoboxes*.

homeotic   a homeotic mutation, as first defined in 1894 by William Bateson, is one that transforms the structure of one region of the body into that of another. An example is the *Ultrabithorax* mutation in *Drosophila*, which transforms the halteres into normal wings. Genes whose mutation causes such changes contain a homeobox; today, all genes that contain homeoboxes are called homeotic genes.

homologous pairs   the pairs of chromosomes of similar kind that occur in most cells, except egg and sperm cells which contain only one chromosome of each kind.

homozygous and **heterozygous**   an animal or plant is homozygous for a particular unit character when the pair of genes that control the character are identical in function (e.g., if each gene contributes equally to the formation of a red-flowered plant). If they differ in function (e.g. if one gene in a red-flowered plant is inactive), the plant is heterozygous.

Hox **genes**   a subgroup of homeotic genes involved in determining differences in structure along the anterior-posterior axis of an animal.

hybridization   the hybridization of one DNA and one RNA molecule is their formation of a double-helical structure stabilized by noncovalent bonding between adenine and uracil and between guanine and cytosine. To hybridize, the two molecules must have complementary sequences of nucleotides, that is, those that allow bonding between each successive pair of nucleotides.

imaginal disc   groups of cells within the larva of *Drosophila* and other insects from which the adult fly is largely formed within the pupa.

induction   a change in cell composition resulting from the receipt of a signal molecule at the cell surface.

intron   sequence of nucleotides within a gene, which, after transcription into RNA, must be excised during the formation of messenger RNA.

invagination   inward movement of a sheet of cells.

larva   free-living stage in the development of some animals interposed between embryo and adult; for example, caterpillars are larvae of butterflies.

linkage groups   different genes that tend to pass to the same daughter cell because they lie on the same chromosome.

maternal effect mutation   a change in the structure of an animal that can arise only from a mutation in a gene of its mother; this usually acts by altering the structure of the unfertilized egg.

meiosis   two nuclear divisions in the formation of egg and sperm cells that halve the number of chromosomes per cell and so compensate for the doubling of chromosome number at fertilization.

mesoderm   see *ectoderm*.

messenger RNA   RNA transcribed from a gene that is translated into a protein.

microtubules and **actin filaments**   main components of the fibrous cytoskeleton within cells that control cell shape and movement.

mitosis   the normal cell and nuclear division in which adhering daughter chromosomes (chromatids), formed by the earlier replication of each chromosome, separate into the two new cells.

molecules   particles formed when atoms are joined by covalent (or sometimes ionic) bonds; they often associate further by three kinds of weaker noncovalent bonds.

morphogen   a compound that induces a different change within cells at different concentrations.

morphogenesis   the formation of tissue, organ, and body shape and structure.

mutation   in a gene, an alteration in its nucleotide sequence, which either inactivates it or causes it to form RNA, and hence protein, of altered structure.

neural crest   the border of newly induced nervous tissue in vertebrate embryos from which cells migrate to form tissues in other regions of the body.

noncovalent bonds   weak bonds of three kinds (hydrogen bonds, ionic bonds, and van der Waals attractions) that stabilize conformations and also guide numerous essential interactions between cell molecules.

notochord   a flexible rod derived from mesoderm cells, running the length of the body of members of the phylum Chordata of which vertebrates including man are the most familiar; present in most vertebrates only in the early embryo and later becoming part of the vertebral column.

nuclease   an enzyme that breaks a polynucleotide chain.

nucleosomes   particles composed of two molecules each of four kinds of histones. A fifth histone (H_1) links adjacent nucleosomes. The DNA helix in a chromosome is wound around a series of nucleosomes.

nucleotide   see *DNA* and *RNA*.

parasegment   term used for 14 temporary subdivisions along the anterior-posterior axis of the early *Drosophila* embryo. Cells at the front of each parasegment give rise to the rear of one segment, whereas those at the rear give rise to the front of the following segment.

polysaccharides   large molecules formed by the covalent bonding of sugar molecules into long straight or branching chains.

promoter   nucleotide sequence, adjacent to a gene, to which RNA polymerase and certain transcription factors bind.

proteins   compounds that are built up within the cell by covalent bonding into long chains of 20 amino acids. Proteins differ from one another in the number of amino acids in the chain and in the order in which the different amino acids are arranged. During its formation, the protein usually coils into its own characteristic three-dimensional structure or conformation. Although protein molecules are very large, samples of proteins can nevertheless be isolated pure, that is, composed of molecules of identical structure.

pupa   inert structure formed by a larva when it is transforming into an adult.

pure   for chemists, pure has a precise and limited meaning, that is, a pure sample of a compound is one in which all molecules are identical.

recessive   see *dominant*.

RNA   ribonucleic acid. RNA molecules have a very similar structure to that of DNA molecules, being built up within cells from four small nucleotides named adenylic acid, uridylic acid, guanylic acid, and cytidylic acid (usually abbreviated to A, U, G, and C). They can, but usually do not, occur in complementary pairs like DNA molecules.

RNA polymerase   an enzyme that catalyzes the transcription of DNA into RNA.

segment   repeating subdivisions along the length of many animals. Both the larva and adult *Drosophila* have 14 segments, but there is no complete subdivision of the body as the word might suggest. Some segments are easily seen, such as the three thoracic segments of the adult fly, each of which carries a pair of legs. Others, such as those in the head, are less obvious.

somites   groups of mesoderm cells that lie along each side of the embryonic notochord and give rise to muscles.

tissue   an assembly of similar cells that contributes to the structure of the organs of which the body is composed.

transcription   the process in which a gene guides the formation of a molecule of RNA whose nucleotide sequence is complementary to that of one of the two complementary DNA molecules (or "strands") of the gene.

transcription factors   proteins that bind to nucleotide sequences adjacent to a gene and regulate its transcription by RNA polymerase. Specific transcription factors regulate only one or a few genes and so initiate differences in cell composition during development.

translation   process in which a particular nucleotide sequence in messenger RNA guides the formation by cell enzymes of a protein molecule with a particular amino acid sequence.

vegetal hemisphere   the half of its egg cell in which nutritive yellow granules are concentrated.

vegetal pole   the center of the vegetal hemisphere of the egg cell.

ventral   see *dorsal-ventral axis*.

vitalism   the belief by scientists, who came to be known as "vitalists," that the activities of living organisms cannot be explained solely by the interaction of their component atoms and molecules by known chemical and physical forces.

wild-type   the familiar form of an animal or plant in contrast to mutant variants.

zygote   the fertilized egg cell.

INDEX

T - #0378 - 101024 - C176 - 229/152/10 - PB - 9781560329367 - Gloss Lamination